A Leap in Science and a Step of ...
Seeking God for the Scientifically Curious

© 2020 Daniel S. Zachary

All rights reserved. No part of this book may be duplicated, copied, translated, reproduced, or stored mechanically or electronically without specific, written permission of the author.

All Scripture quotations, unless otherwise indicated, are taken from the New Revised Standard Version Bible, copyright © 1989 the Division of Christian Education of the National Council of the Churches of Christ in the United States of America. Used by permission. All right reserved.

ISBN: 9781654195632. Printed in the United States.

All Scripture quotations, unless indicated, are taken from the New Revised Standard Version Bible (NRSV), copyright © 1989 the Division of Christian Education of the National Council of the Churches of Christ in the United States of America. Used by permission. All rights reserved.

For more information on this subject matter, please keep current by going to the Faith & Science website, https://dszachary.com. To keep costs reasonable, the printed version of this book is in black and white (the electronic version is in color).

Cover design by Andrei Publico, Manila, Philippines
Chapter logo design by Ulysse Zachary

Some chapters contain figures by the author, including:
Ch 1: Figures of Knowledge
Ch 3: The relative strengths of the force fundamental forces
Ch 3: A schematic of energy excitation
Appendix: Two concepts of time for our universe; Skaters
Also, the second figure in Chapter 3 is partly done by Calypso Zachary.
All other figures are credited to their sources.

Endorsements for *A Leap in Science and a Step of Faith*

A Leap in Science and a Step of Faith by Daniel Zachary is a scientist's attempt to take a fresh look at where science and Christianity intersect. I would say that Daniel succeeds greatly in his purpose here. He asks us to take a new look at questions such as the nature of reality, the role of chance, and the relationship between scientific and biblical views of the world. This book will be helpful to the nonscientist who wants to understand how a person who begins with evidence to form theories about reality can come to faith in the Bible as the inspired Word of God.

– Dr. John Oakes, professor of chemistry and physics, Bible teacher, and author

A Leap in Science and a Step of Faith is a thoughtful exploration of the complex relationship between spiritual belief and scientific inquiry. It features clear explanations and examples, making it an accessible text for readers without advanced training in science or theology. Friendly to believers and nonbelievers alike, this book weaves biblical teachings and scientific principles together with stories from Zachary's own personal journey.

– Jade Olson, Ph.D., Lecturer in Communication, University of Maryland, College Park (nonreligious affiliation)

For many, the Bible is seen through the lens of science, and this can undermine its true purpose. *A Leap in Science and a Step of Faith* takes a different approach, highlighting science for what it does so well—revealing the amazing facts about the universe.

These facts ultimately show that it takes a lot of faith to believe that we are here by pure chance. With a similar fact-seeking approach, Dan Zachary highlights numerous fascinating aspects of the Bible, exploring the evidence for its validity and how both testaments point to its central figure, Jesus Christ.

– Dr. Douglas Jacoby, MTS, DMin, International Bible teacher and adjunct professor of theology at Lincoln Christian University

Contents

Preface iv
Part I – The Paths We Take 1
   Chapter 1 – Which Path? 1
   Chapter 2 – A Leap or a Step? 19
   Chapter 3 – Tracing Earlier Steps 33
Part II – A Leap in Science… 51
   Chapter 4 – Our Fine-Tuned Universe 52
   Chapter 5 – Our Earth: A Unique Paradise or One of Many? 74
   Chapter 6 – First Life 118
Part III – …and a Step of Faith 142
   Chapter 7 – The Bible Through the Lens of Science 143
   Chapter 8 – The Bible Through the Lens of History 159
   Chapter 9 – God of the Gaps or Gaps about God? 176

Epilogue – My Own Journey 213
Appendix 1 – Faith: Biblical and Scientific – Two Sides of the Same Coin? 219
Appendix 2 – "In the Beginning" 222
Appendix 3 – Scientific Concordism and Skating 229
Appendix 4 – A Proof of Chronology: The Dead Sea Scrolls 234
Appendix 5 – Messianic Prophecies 239
Acknowledgements 247
Bibliography 249

Preface

My spiritual journey began some 36 years ago. Long before the days when I made any significant attempt to put the Bible into practice, I was already convinced there was something more to this universe than met the eye. As a student in astronomy and physics who had some knowledge of the Bible, I was intrigued by discussions that involved God and science. When I got tired of studying, I would take breaks, usually around midnight, and meet up with my friend Serge, a classmate in physics and a resident down the hall. We both loved debating whether God exists and how he might have intervened in our universe.

At the time, I knew just enough of the Bible to offer a nice prepackaged but rather unsubstantiated "God" presentation to Serge. Serge would respond in kind with all sorts of theoretical suggestions ranging from general relativity to quantum physics. We both listened to each other, and sometimes other classmates, most of whom were studying science, engineering, and the like, would join in and add flavor to the discussion. My "God-based" arguments made sense to me; after all, I had read through the Bible twice already and thought I knew enough to offer my opinion. Applying my twentieth-century mindset to this ancient book, I extracted what I thought was Bible-based science. As you might have guessed, my friend was not impressed.

Serge was from Eastern Europe, a brilliant student, training to be a theoretical physicist; he was also a hardened atheist. When I think back to some of the arguments I presented to him, I am surprised that he listened to me. I spoke with misplaced authority based on my limited knowledge of the fundamental sciences and a seriously misinformed view of the Bible. I am grateful that we remained good friends throughout our studies until we parted ways after graduation.

A few years after those midnight chats with Serge, I got serious about putting the Bible into practice. Soon after, I became interested in the history of the Bible and learning about how it came together. During this time, I pursued a career as an astrophysicist and later as a nuclear scientist. I slowly became better at listening to people with questions and learned to have more productive discussions with scientists, atheists, and the average Bible skeptic.

After graduating, in the mid-1990s, I moved to Geneva Switzerland to pursue my postdoctoral work. I continued to have these "midnight discussions," now afternoon coffee breaks, with friends and colleagues at the European Laboratory for Particle Physics (CERN). I began taking notes from these conversations, which gave me a window into the patterns in people's views. The foundation of this book came from piecing together these accounts, bringing in the most up-to-date science, and gathering the most recent knowledge of Bible evidence.

Aim and layout of the book

The book has three parts. The first part lays out the landscape of views related to science and the Bible. We will look at some of the great scientists from the Renaissance period through the twentieth century. We will explore why, for some of them, their faith was strengthened in their scientific pursuits, while the faith of others was weakened. In the second part of the book, we'll explore the sciences of origins, both physical and biological. In the third part of the book we'll turn to exploring evidence surrounding the Bible. When we do this, we won't abandon our reason and senses. As we spend some time looking at the evidence supporting the veracity of the Bible and its claims, the version I will use is the New Revised Standard Version (NRSV). This version is considered one of the most accurate English versions available (see info on the inside cover of this book).

Finally, we will address skepticism about the Bible by working through spiritual issues via the lens of science and math. We will raise a few questions at the end of each chapter.

Fast tracking through this book

Numerous references fill this book; these are for the curious student. Some readers may choose to skip the endnotes for a briefer read; others may want to read every note to understand the backstory.

Who might appreciate this book?

I wrote this book for Christians, especially Christians who are facing doubts and questions about how to reconcile their faith with science. If you are scientifically inclined or spiritually curious, you may appreciate this material. Some of the concepts in this book may be difficult and may even push us to our "quantitative thinking limit," but most of the book provides illustrations for the reader. This book may also be helpful as a tool when discussing faith with a nonbeliever, especially one who is inclined toward the sciences, history, and evidence-based thought.

Final thoughts before we start

Hopefully, this book will help you to ask questions. As a teacher, I like to ask questions. I have learned that the trick is to ask the right kind, ones that get someone to think, not necessarily to work toward a solution, but to examine different viewpoints. As ironic as it sounds, the best questions raise even more questions.

I like the advice of Albert Einstein: "The important thing is not to stop questioning." He was right. And let's also consider the evidence we discover—it may change our way of thinking. The famous atheist and philosopher turned believer, Anthony Flew, once said, "You must go wherever the evidence leads."

Dan Zachary, June 28, 2020

Part I – The Paths We Take

Chapter 1 – Which Path?

"There was nowhere to go but everywhere." – Jack Kerouac

I was walking through the Alps in late winter, along with a friend. We were on snowshoes and on our way to the wedding ceremony of another friend who loved the mountains and remote places. We started off snowshoeing on one side of a ridge and hiked over it. The idea was to descend into a valley where a mountain lodge was situated and where we would find the wedding party. Once we reached the peak of the ridge, we looked down into the valley and searched for a path; dusk was approaching. We could see footprints in the snow going off, more or less, in the direction of the lodge, but the paths diverged, leaving us to decide which one to take. We both knew that, given the isolation of this valley and the fact that steep slopes surrounded it, any path should lead to our destination. But we hesitated and looked at each other. Which one?

Finding your path

We chose a path and managed to find the lodge, joining the rest of the party with other friends who had arrived earlier by taking the road that led up from the other side of the valley. Though most took the easy way to the lodge, the two of us had fun and arrived at the door just as the moon began to rise over the mountains.

Finding your path toward God is a bit like finding the right path on this mountain slope. Finding the correct path is part of the Christian experience; this is true when it comes to seeking God, but it is also true when it comes to balancing what we observe with what our faith tells us. For the scientifically curious, this is especially difficult. On the one hand, it's not hard for most Christians to look up at the sky on a moonless night and be amazed at the creation. Perhaps we think about God at the beach, or maybe we enjoy a psalm or the comforting words of Jesus as we walk through the woods. But when we dig further into the Bible, the messiness (or challenge) begins. For example, it can be confusing when we begin to read through the first sections of the Bible, especially the book of Genesis, and try to reconcile our faith with what we have learned from science. Untangling Bible passages that refer to the creation is a bit like traveling down the mountain paths mentioned above—perhaps all paths lead to the same lodge, but they diverge along the way. Maybe some paths go more or less directly toward the lodge, while others meander through the forest. In any case, it's nice to have some kind of map on these journeys.

The central message of Christianity is clear: Christians understand there's only one door, or gateway, through which we must pass to know God, and Jesus is that gateway. And as challenging as this concept is for believers, it is even more difficult for nonbelievers. I would also argue that the science vs. Bible conflict we face today makes the challenge even more daunting.

Some Christians have grown up in a Bible culture (as did the author). But even with a religious background, it can be very challenging to become a Christian—the path is snowy, meandering, and sometimes very long. Imagine now someone with no Bible background and how they would fare on that same journey. Now, imagine the same person facing a biblical creation story and bringing into that study only what education has presented to them from a scientific perspective. The person risks remaining paralyzed on the proverbial mountain ridge as they stare down the snowy paths. The goal of getting to the lodge may get lost before they even start the trek; this is where Christians who are hoping to help their science-oriented friends need to be careful that they present their case in a way their friends can grasp.

Accommodating the paths

For generations, most of the Zachary family lived in and around Zachary, Louisiana,[1] a town with a culture and heritage that was, for better or worse, overflowing with "religion." For someone growing up in the religious community of Zachary, they will most likely have heard a narrative of a conservative creation story, including Adam and Eve, the Garden of Eden, and the entire thread of Genesis characters. Whether or not that story coincides with science would not necessarily be problematic, since the culture supports this traditional creation story.

Now imagine someone with no Christian background, maybe someone from Eastern Europe or China, who is transported to the town of Zachary. Imagine subjecting that person to a traditional creation message—Adam and Eve, etc. What would that person think? Would they be comfortable with these stories? Most likely, they would need biblical accommodation.

Accommodation is a way to convey an important message so that it's understandable to the listeners. Usually, accommodation means simplifying the message, but occasionally, depending on one's background, an oversimplified biblical creation message can be just as awkward or unsuitable.

Jesus was a master at presenting an accommodating message for his immediate listeners. His task was to present the message of a sovereign, omnipotent, omnipresent God to shepherds and fishermen. He did this by using parables containing sheep, goats, yeast, grain, seeds, plants, fish, and other objects that people could understand. What would be the metaphors Jesus would use in our developed world? Might they not contain allusions to smartphones, social media, and COVID-19?

Jesus was not the only master at creating an accommodating message. C.S. Lewis was perhaps one of the most accommodating authors of the twentieth century and had a certain wisdom in reaching the nonreligious reader. This British author is most known for his children's stories about Christianity, the *Chronicles of Narnia*; however, one of his other works, *Mere Christianity*,[2] is unequaled in the way it conveys a no-nonsense, basic, or "mere" message of Christianity. The book relies on common sense and logic, and addresses moral principles as it helps us to understand the most rudimentary part of Christianity, the concepts of forgiveness and the Incarnation. The entire narrative of *Mere Christianity* takes an accommodating, thoughtful, and humorous approach. C.S. Lewis' work is useful for the person who needs to first hear logic and candid thought as opposed to religious traditions and stories. Indeed, Lewis does not mention a single passage from the Bible, although the entire work is about Christianity—it's accommodating.

A Christian may need to "accommodate" to reach those with a scientific background; this does not mean altering science to fit the Bible, or even modifying the Bible to fit science; it merely means putting the message in a package that can be understood by the listener. This kind of message, though gaining some popularity in recent years, is still lacking for many would-be God-seekers.

The two books of God

According to the Bible, God communicates or reveals himself to us in many ways. For this discussion, we will consider two of these ways, or the "two books" of God: nature (general) revelation and the Bible (special) revelation.[3]

As we mentioned above, the central message of Jesus was very accommodating, albeit challenging. Science, on the other hand, is quite different. It can be very accommodating in revealing God in specific contexts, for example, when we reflect on God under a clear summer night sky, but it can also be very unaccommodating. Though nature provides easy first access to understanding God, as soon as one studies fundamental science in any depth, the easy, accommodating message becomes blurry. Unless you are prepared to spend significant time sharpening your knowledge in research, most likely you will not see how new revelations in science help you to understand God. If you do, you may have a difficult time communicating that to others.

Specialists in my field of science (physics) spend their entire careers building and supporting arguments to describe nature. The technical details of the fundamental sciences can completely obscure the notion of God to all but a small number of the people who invest their time in these investigations. Even the description of reality becomes difficult to grasp, let alone the precise mathematics used to quantify nature.

These technical challenges inherent in science don't imply that they are useless in understanding God. They only mean that the two books of God are not interchangeable in terms of accommodation. Some have taken the approach that the two books are connected. In other words, not only does each book point toward God, but nature and Scripture can be harmonized—the Bible points to nature and nature points to the Bible. This harmonization has been referred to as *scientific concordism*.[4]

In this book, we take a different view. We propose that nature and the Bible, *each on their own merit*, handily point toward God. *The more one understands nature at its deepest levels, or the more one understands the Bible in its proper context, the clearer the path toward God becomes. Though nature and the Bible each have their merit, either one can help someone find God. Nature is an appropriate place to start the journey; it is the Bible that contains the specific instructions for the seeker to complete that journey.*

Christians following different paths

We describe four general paths Christians can follow as they reason through the science and the biblical creation issues. The four views are generally divided along the lines of Bible interpretation and include: 1) Young Earth Creationism, 2) Intelligent Design, 3) Progressive Creationism, and 4) Evolutionary Creationism.[5] To complete the list, there are two other views that don't fall under Christian theology, and we will include them in the following section.

Young Earth Creationism

Young earth creationism (YEC) assumes a *literal view* of the Bible. It holds to the concept that the universe, the earth, and all lifeforms were created in a supernatural act of God in the recent past. This view sees the creation occurring in six days, or a total of 6×24 hours during the creation week account as recorded in Genesis 1.[6] This event took place approximately 6,000 to 10,000 years ago. More precisely, James Ussher, a seventeenth-century Archbishop of Ireland, developed a chronology by summing up the ages of the fathers at the time of the birth of their sons for the entire lineage between Adam and Christ.[7] He calculated the date of the creation to have occurred at nightfall on 22 October 4004 BCE.

Since young earth creationists view the creation as a miraculous event, they do not accept *macroevolution*, a process that requires millions to hundreds of millions of years.[8] This view does accept small evolutionary changes (*microevolution*), but any changes are minor given the timescale of 10,000 years or less. For example, different types of dogs may have evolved or changed by breeding over this range of time, but YEC doesn't allow enough time for dogs and wolves to evolve to their separate species.[9] Supporters of this view believe that all species that ever existed first appeared during the miraculous creation week. This view implies that dinosaurs and other extinct species must have therefore lived alongside humans but somehow disappeared, perhaps in the great flood described in Genesis 6.[10]

From a literal perspective, YEC is adherent to the Bible. On the other hand, many scholars disagree with YEC on the grounds that it does not consider the context of the ancient writings.[11] Furthermore, science does not support the concept of a young earth.[12]

Intelligent Design

Intelligent design (ID), argues for the existence of God by presenting what it believes to be "an evidence-based scientific theory about life's origins."[13] Proponents of this group see God intervening in miraculous ways from time to time and in different places throughout the creation process. Intelligent design adherents have an "old earth" concept, understanding that both the earth and universe are very old, with ages consistent with scientific observation. However, ID departs with mainstream science typically in the context of biology, life systems, and evolution. Two fundamental concepts of this view include *irreducible complexity*[14] and *specified complexity*.[15] Irreducible complexity, for example, implies that evolution cannot explain the complexity of certain biological systems via a succession of small modifications. Even the most basic unit of life, the cell, is considered far too complex for natural processes to have created it, and therefore, an intelligent being (God) must have intervened.[16] This intervention would have occurred at specific points in the evolutionary track to "assist" macro changes or create new species. Some view the appearance of new species or "kinds" as interpreted in the biblical creation account as a scientific concordism, where the Bible corresponds with modern science.

Intelligent design advocates take a *literary view*[17] of the Genesis creation account or a day/age view, accepting the chronology of the creation account of Genesis 1, but viewing each day representing vast periods of time.[18] This group presents a defendable position in terms of the physical creation by fully accepting the scientifically supported old-earth concept that includes the evidence of the geology of an ancient world and the fossil evidence showing extinct and extremely old prehistoric life. What ID does not accept is a natural, or random process in major evolutionary events, for example, the mutations from reptiles to birds, and the concept of *common descent*.[19] Intelligent design supporters view the gaps in fossil evidence at these junctures as evidence that a miraculous intervention took place. These same gaps support the concept of *God of the gaps*,[20] a source for some criticism in the case where scientific knowledge has filled these gaps from time to time.

Progressive Creationism
  Progressive creationism (PC) accepts the implications of evolution and yet holds the theological position that God is involved in the process. Advocates of PC also accept that the earth and universe are both very old, consistent with observations from science. PC adherents "may lean toward a more literary view of the Genesis creation account or they may take the view that is based on scientific concordism, that there is at least a general scientific truth in the chronology, if not the time span, in Genesis."[21]
  Progressive creationists point out that archaeological records show that "a species does not arise gradually by the steady transformation of its ancestors; it appears all at once and 'fully formed.'"[22] Finally, progressive creationists distinguish themselves from intelligent design advocates by accepting the main implications of evolution, including the concept of common descent as a scientifically valid hypothesis. Furthermore, PC is skeptical of the "gap argument," a key feature in intelligent design.[23]
  The strengths of progressive creationism is that it provides for a creation account consistent with a literary view of Genesis 1 and simultaneously consistent with the main components of scientific evolution. God intervenes in subtle ways, guiding the entire process of evolution. Some have criticized PC saying its predictions

concerning evolution are too vague; in its attempt to satisfy both biblical and scientific views of evolution, PC is sometimes not clear on what is supernatural and what is natural in the evolutionary process. In other words, the balance between free will in nature (randomness) and the intervention from God is difficult, if not impossible, to pin down to prove or disprove in this view.

Evolutionary Creationism

Evolutionary creationism (EC) accepts the general theory of evolution from the mainstream science community and views that all species, including humans, evolved from the simplest life forms in natural processes. All concepts of modern evolutionary theory, *neo-Darwinianism*,[24] are also accepted by this view, including the concepts of common descent, *natural selection*,[25] *genetic mutation*,[26] and *genetic drift*.[27] Evolutionary creationists naturally accept an old earth view and nearly unanimously reject scientific concordism—EC accepts the theology of Genesis 1–11[28] but not the science or the history of the accounts.[29] On the other hand, EC views God as having the precognition to establish a priori the entire chain of evolution before the creation. In other words, the initial conditions of the universe were established perfectly by God, via the natural (random) process that took place and that would eventually result in a human (or humans) who could have a relationship with him. There would be no need in this view for God's intervention at specified places (gaps) in the evolutionary process, nor would he need to intervene in a progressive way along the chain of evolution. In other words, EC views God to be sovereign enough to have achieved all necessary preset qualities in the universe for us to be here, an a priori condition of noninvolvement from God, at least in terms of unfolding creation over billions of years. Advocates of this view call this a *God-of-no-gaps theology*.[30]

The strength of evolutionary creationism is that it is in complete agreement with the modern views of cosmology, geology, and evolution. EC also fully accepts the theological implications of the Genesis creation account, that humankind is separated from God by sin. A weakness of EC may be its inconsistency in how God interacts with the creation. For example, evolutionary creationists acknowledge that God has intervened in our lives, through the historical record (the

Bible), through Jesus, and through prayer. Before this time, God did not intervene throughout the entire creation process.[31] Also, as a consequence of this view, Adam and Eve never existed; humans were created in the likeness of God, but that came through a natural, evolutionary process and was not imparted by God in a miraculous event.

As a way to summarize the four Christian-based views on evolution and the age of the earth, the evolutionary track for vertebrates (see the figure on the next page) depicts how each group might view how God intervened. The salient distinctions of different views include the frequency of God's intervention. The circles represent points where intervention might have occurred (the ID position). The PC view suggests that God "oversees" evolution while EC proposes that God intervened, but only at the very beginning (of time). YEC posits that all creation occurred very recently (very near the top of this diagram).

Another underlying distinction between the views is the reliance on or rejection of scientific concordism in the approach to harmonize (or not harmonize) the Bible creation account with science. Scientific concordism is prevalent among ID and PC adherents but virtually absent in the EC view. On the other hand, all four Christian views accept the concept of sin, the deity of Jesus Christ, and the physical resurrection.[32]

Apart from those who are undecided, the above groups represent Christans' views on the creation/evolution process. We also provide two non-Christian views (see below). It is helpful to understand our own view, and it is helpful to understand the views of others; it behooves each of us to remain congenial as we discuss our view with those not in "our camp." Christians' views of evolution and the details of where and when God intervened is not a central dogma in terms of faith. On the other hand, what is important is that we can answer our friends who have questions about Christianity and present our view on creation if our audience so demands. A key concept can be lifted from the pages of 1 Peter:

> Always be ready to make your defense to anyone who demands from you an accounting for the hope that is in you; yet do it with gentleness and reverence. (1 Peter 3:15–16)

10 | The paths we take

We should also remember that our path is dynamic; we may change views as we go through life, somewhat like changing paths on the snowy mountain slopes in the example above. Changing paths is not necessarily a bad thing, but it reflects the challenge that all groups have in answering all questions, physical and spiritual: no view provides all the answers, and no view is entirely wrong.

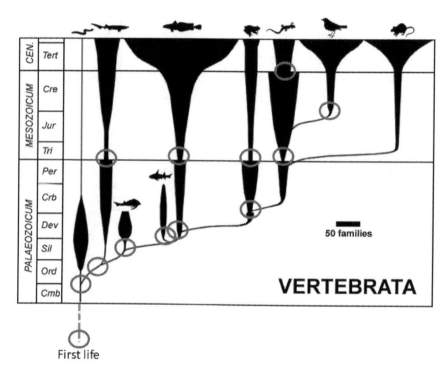

The evolutionary track for vertebrates; see text, above, for description (Source, Public Commons; Adapted from P. Bøckman)

Other views and summary
Two non-Christian views also describe creation.

Deistic evolution
Deistic evolution (DE) is a position that accepts all scientific evidence for evolution, including an old age for the earth and the

Which path? | 11

universe. A supreme being or force initiated the entire creation process, but that being is detached. This view rejects the notion of a personal God. The strength of this view is that it accepts the complete scientific account of evolution. On the other hand, DE rejects the tenets of Christianity; neither the Incarnation nor the physical resurrection of Jesus is accepted.

Atheistic Evolution

Atheistic evolution (AE) is perhaps the easiest to describe, as it views the entire creation as resulting from natural or random processes. No God or deity is responsible for the initiation of the creation, and, of course, evolution is based entirely on natural processes.[33]

	Age of Universe*	Evolution of Life	Interpretation of Genesis 1–11	Origin of Humanity	Scientific Concordism	Examples
YEC	Young	Limited Micro-Evolution	Strict Literalism Day=24 hours	Adam & Eve **	Yes	Ken Ham – Answers in Genesis
IC	Old	Only Micro-Evolution	General Literalism Day/age view	Adam & Eve**	Yes	Stephen Meyer – Discovery Institute
PC	Old	Only Micro-Evolution	General Literalism Day/age view	Adam & Eve**	Yes	Hugh Ross – Reasons to Believe
EC	Old	Yes	Spiritual truths Ancient science Ancient poetry	Humanity** evolved	No	Francis Collins – *The Language of God*
DE	Old	Yes	Fairy tale	Humanity*** evolved	No	Charles Darwin – *The Origins of Species*
AE	Old	Yes	Fairy tale	Humanity*** evolved	No	Richard Dawkins – *The God Delusion*

Top portion represents groups with Christian theology – including the Incarnation and physical resurrection
* Young ≈ 6000 – 10,000 years; Old, Earth is ≈ 4.6 billion years old; the universe ≈ 13.8 billion years old
** Notion of being created in the image of God/accepts human sin
*** Rejects being created in the image of God/rejects human sin

Views of the origins of the universe and life (Source: Adapted from D. Lamoureux, Evolution: Scripture and Nature Say Yes, *2016)*

The table summarizes the main views.[34] In full disclosure, if forced to choose, the author's view overlaps with PC and EC groups and is nonconcordist. The presentation in Chapter 7 speaks to this view.[35] That being said, I would now like to provide a different approach that speaks to all groups.

Science-theology dualistic view

There is no doubt that the evolutionary/creation views greatly vary, and this can be a source of frustration for the Christian and of confusion for someone attempting to make sense of the Bible and Christianity. Before we leave this chapter, I will provide you with a thought.

When I was a junior in college, one mandatory course for a physics major was a class called Junior Lab. Students took this course in their third year of undergraduate studies, but there was nothing "junior" about it. Two Nobel laureates ran the breakout sessions in these labs—one of them was one of the friendliest instructors I have ever experienced, the other was rather grumpy. Both were brilliant. It was in the breakout session with the grumpy professor that we explored some of the classic experiments that generated the "leap in science" in the first part of the twentieth century. One such experiment was to perform the famous wave–particle duality experiment.[36] For our discussion here, details of this experiment are less important than the result. What the experiment showed is that at small scales, we observe that matter acts both as solid particles and as waves. In other words, our universe is dualistic—matter can be two things at the same time, a remarkable finding! Though physicists have accepted this concept for more than a century, the concept remains as bizarre as ever. How can matter be two things at once? But this is just the way matter behaves, and both views are needed to describe our universe.

So what does this laboratory experience have to do with our view on creationism? If wave–particle duality is valid for physical reality, I would argue that there is a certain duality concerning the creation picture—it is both physical and theological. Perhaps the Genesis account is entirely accurate from a theological perspective, and the description of the physical creation is entirely correct from a scientific perspective; they go hand-in-hand, and each remains pertinent in its own account of how reality unfolded. One can't describe the other, but together they provide the full picture of our reality. This picture does not obviate the effort we should make to pursue proper biblical *hermeneutics*,[37] any more than it nullifies careful science. Still, it should give us room for reflection about our view of the creation. Understandably, a dualistic view may seem like a copout to some as we allow for all four Christian views described above (YEC, ID, PC, EC), but there is more to this than mere acceptance for everyone.

A purely theological view of Genesis 1 should not influence the scientific view, nor should the purely scientific view influence the theological view. Given these two extremes, we do not expect YEC to have much to say concerning the physical creation as it presents the theological aspects of Genesis 1. Similarly, we do not expect EC to have much to say concerning the theology of Genesis 1 as it offers a presentation on scientific evolution. Also, it is reasonable to view ID and PC as representing intermediate positions between these two extremes.

In terms of an illustration, on one end of the creation spectrum are miraculous events, and at the other end is the physical creation; both sequences of events unfolded, but like particles and waves, only one can be viewed at a time, though they coexist. Similar to the wave–particle duality of nature, there are times when nature only has one property. A billiard ball hits another billiard ball on a pool table, and we know what happens after the collision using particle properties of matter. No wave description would be appropriate here. The beauty of a rainbow is understood only using wave properties of light. No particle description of light would make sense here. There are subtle overlapping situations that we can produce in the laboratory—we understand that if a billiard ball became small enough, we could describe it as a wave; if light became energetic enough, we could describe it as a particle. Physicists accept both realities of nature—particles and waves. Is the creation story perhaps analogous?

There is a theological counterpart of duality that comes directly from Jesus and his famous encounter with the Pharisees and the woman caught in adultery.

> Jesus went to the Mount of Olives. Early in the morning he came again to the temple. All the people came to him and he sat down and began to teach them. The scribes and the Pharisees brought a woman who had been caught in adultery; and making her stand before all of them, they said to him, "Teacher, this woman was caught in the very act of committing adultery. Now in the law Moses commanded us to stone such women. Now what do you say?" They said this to test him, so that they might have some charge to bring

against him. Jesus bent down and wrote with his finger on the ground. When they kept on questioning him, he straightened up and said to them, "Let anyone among you who is without sin be the first to throw a stone at her." And once again he bent down and wrote on the ground. When they heard it, they went away, one by one, beginning with the elders; and Jesus was left alone with the woman standing before him. Jesus straightened up and said to her, "Woman, where are they? Has no one condemned you?" She said, "No one, sir." And Jesus said, "Neither do I condemn you. Go your way, and from now on do not sin again." (John 8:1–11)

Perhaps one of the most touching stories in the Gospels is woven into this unusual duality as the grace offered by Jesus is in juxtaposition with sin and its punishment as required by the law (Exodus 20:14). The Pharisees were justified in making their accusation and concluding that she should be stoned to death, but it was Jesus, by his nature, who provided the dual picture (more on this in Chapter 9).[38]

Jesus demonstrated the duality of sin/punishment and grace/forgiveness, but he offers more than that to our creationism view. Jesus' life (for evidence in history, see Chapter 8) reveals the physical revelation of God, and his resurrection (for evidence in history, see Appendix 4) offers a miraculous presentation. The two ends of the spectrum (physical and miraculous) converge in one life, and we can review that life by means of scientific investigation.

The creation is indeed subtle in many ways, and we will see this demonstrated both in nature (Part II) and in the Scriptures (Part III).[39] Though we will study these two topics separately, together they make the whole. The universe is subtle; God is too.

Chapter summary and questions

The Christian community is composed of several camps in terms of the creation story, but the common theme of Jesus' life ties these groups together. Maybe all these groups represent different aspects of the whole, and together they help us understand God and his creation.

Questions to ponder:
- What path are you on, YEC, ID, PC, EC, DE, NE, or other? If so, why?
- Have you ever changed paths (camps)? Why?
- What would you have to see or learn for your view to change? Or do you believe your views are unchangeable?
- Consider someone who was able to accept the position of the evolutionary creationist, but could not accept any of the other positions. Would you welcome that person into your fellowship? Why or why not?

NOTES
[1] Until one of the Zachary family, my grandfather, moved from Lousiana to the Motor City for work in 1925.
[2] Robert Fulford, "How C.S. Lewis made Christianity seem like common sense," National Post, December 21, 2015, https://nationalpost.com/entertainment/books/how-c-s-lewis-made-christianity-seem-like-common-sense.
[3] More will be said on this topic in Chapter 7.
[4] Scientific concordism refers to the position that the teaching of the Bible on the natural world, properly interpreted, will agree with the teaching of science (when it properly understands the data), and may in fact supplement science. *Dictionary of Christianity and Science*, edited by Paul Copan, Tremper Longman III, Christopher L. Reese, and Michael G. Strauss (Grand Rapids, MI: Zondervan, 2017).
[5] There are other ways (theories) to classify these groups. We use categories that have been more popular in recent years. For a good overview of the four Christian views, see the essay by John Oakes, "Four Christian Views of Evolution: An Essay," June 6, 2010, https://evidenceforchristianity.org/four-christian-views-of-evolution-an-essay/. We also note that these groups cut across denominational lines.
[6] More will be said on this topic in Chapter 7.
[7] James Ussher, *Annales veteris testamenti, a prima mundi origine deducti* (Annals of the Old Testament, deduced from the first origins of the world), which appeared in 1650, and its continuation, *Annalium pars posterior* (Annals posterior), published in 1654.
[8] Macroevolution: large-scale evolution encompassing the greatest trends and transformations occurring over millions to hundreds of millions of years.
[9] G. Larson and D.G. Bradley, "How Much Is That in Dog Years? The Advent of Canine Population Genomics," 2014, https://journals.plos.org/plosgenetics/article?id=10.1371/journal.pgen.1004093.

[10] There are forms of YEC that attempt to explain or reconcile a young earth (as interpreted in Genesis 1) with the apparent very old earth. Gap creationism proports that a large "gap" of time existed between Genesis 1:1 and Genesis 1:2 during which the creation was destroyed, and that Genesis 1:2–31 describes a re-creation. We note that there is no direct biblical support for the gap theory, nor is there any scientific evidence for it.

[11] More will be said on this topic in Chapter 7.

[12] More will be said about the age of the universe and life in Chapters 5 and 6.

[13] Stephen C. Meyer, "Not by chance," *National Post* (Don Mills, Ontario: CanWest MediaWorks Publications Inc., December 1, 2005). Others view this position as a pseudoscience since the statements or beliefs of ID are claimed to be both scientific and factual but are nonfalsifiable and therefore incompatible with the scientific method.

[14] Irreducible complexity (IC) is a central component to intelligent design. The idea is that certain biological systems (e.g., the cell) cannot evolve by successive small modifications to preexisting functional systems through natural selection.

[15] A property that singles out patterns that are both specified and complex. For example, "Living organisms are distinguished by their specified complexity. Crystals are usually taken as the prototypes of simple well-specified structures, because they consist of a very large number of identical molecules packed together in a uniform way. Lumps of granite or random mixtures of polymers are examples of structures that are complex but not specified. The crystals fail to qualify as living because they lack complexity; the mixtures of polymers fail to qualify because they lack specificity." Leslie Orgel, *The Origins of Life* (London: Chapman & Hall, 1973), p. 189.

[16] More will be said on this topic, especially in relationship to the origin of life, abiogenesis, in Chapter 6.

[17] The literary view, as it pertains to the Genesis account, "accepts the theological implications of the creation account, but not what some see as the apparent scientific content of the creation story. In other words, the Genesis creation account is to be seen strictly for its literary content as poetry. It describes theological themes, but not chronology. Whether the "events" occurring on the fifth day happened after the "events" on the second day is not germane to what the inspired author is trying to tell his audience. The days are not ages at all. They are categories of what God did." Oakes, "Four Christian Views of Evolution," note 2.

[18] Oakes, "Four Christian Views of Evolution."

[19] Common descent is a concept of evolution that views one species as the ancestor of two or more species later in time. It's understood that the full application of this idea implies that there existed a last universal common ancestor (LUCA) of all life on earth, a central assumption of modern evolutionary theory.

[20] "God of the gaps" is a theological perspective in which gaps in scientific knowledge are taken to be evidence or proof of God's existence. This position will be discussed more in Chapter 9.

[21] Oakes, "Four Christian Views of Evolution."

[22] Stephen J. Gould, *The Panda's Thumb: More Reflections in Natural History* (New York: W.W. Norton & CO., 1982), page 182.
[23] Oakes, "Four Christian Views of Evolution."
[24] Neo-Darwinism describes the integration of Darwin's theory (evolution by natural selection) with the theory of genetics as proposed by Gregor Mendel.
[25] Natural selection, proposed by Charles Darwin, is the differential survival and reproduction of individual organisms due to the differences in the traits of the organisms (their phenotype).
[26] A gene mutation is a permanent alteration in the DNA sequence that makes up a gene. More on DNA in Chapter 6.
[27] Genetic drift describes the variation of a gene in a population, or more precisely, "the change in the frequency of an existing gene variant (allele) in a population due to random sampling of organisms." Joanna Masel, "Genetic drift," *Current Biology*, Cell Press, 21(20): R837-8, October 2011.
[28] Genesis is divided into two main parts. Genesis 1–11 recounts the prehistory of the Bible (what interests us here) and this includes the creation account; Genesis 12–50 recounts the history of Abraham through the twelve tribes of Isreal and their exile in Egypt.
[29] Though not the focus of this book, this view proposes that the Garden of Eden was not real, nor was there a "fall" of man in the Garden. Contrarily, EC views sin from an individual perspective and not from the onset of a single event. A good article on this topic can be found here: Charles P. Arand, "A travel guide to the evangelical creation debates: What is Evolutionary Creationism?" February 28, 2018, https://concordiatheology.org/2018/02/a-travel-guide-to-the-evangelical-creation-debates-what-is-evolutionary-creationism/.
[30] Oakes, "Four Christian Views of Evolution."
[31] Ibid. John Oakes points out that though this is an inconsistency, it's not necessarily incorrect.
[32] More on this in Chapter 8.
[33] More on this atheistic (naturalism) view in Chapter 2.
[34] Dennis Lamoureux, *Evolution: Scripture and Nature Say Yes* (Grand Rapids, MI: Zondervan, 2016), Figure 6-1.
[35] The biblical creation account (see Chapter 7) does not omit the possibility that God could have miraculously intervened, unfolding a creation in literally 6×24 hours and only a few thousand years ago. God could have created the universe with the appearance of age. Miraculous events (e.g., Jesus turning water into fine wine) demonstrate that instantaneous creation is not beyond God's power. But we know the Bible clearly indicates that God's nature is revealed though creation (e.g., Psalm 19:1; Roman 1:20). It might seem contradictory if God would employ a "deceptive" scheme, creating a universe with only an appearance of billions of years.
[36] A series of experiments were conducted in the early nineteenth century that confirmed wave–particle duality, a concept that is fundamental to quantum mechanics. Every (small) particle or quantum entity may be described as either a particle or a wave, but one description is insufficient to capture the full reality.

[37] Biblical hermeneutics is the study of the principles of interpretation concerning the books of the Bible.

[38] This wasn't the only time Jesus provided a duality. Paying taxes to Caesar is another (Matthew 22:15-22).

[39] Though we will see that some of the numbers supporting the idea of a creator are not subtle at all!

Chapter 2 – A Leap or a Step?

By faith we understand that the worlds were prepared by the word of God, so that what is seen was made from things that are not visible. (Hebrews 11:3)

In one sentence, the Hebrew writer pens what has become a great rift between science and faith. That rift represents a misunderstanding, and unfortunately, it has grown over time. Once, where the science and faith communities drew a line, today there is a chasm. Can these sides ever be reconciled?

The love of science came naturally to me. I was quick to pick up books on astronomy in middle school. The more I learned about the universe, the more curious I became. I'll never forget watching the beautifully orchestrated documentary, *Cosmos: A Personal Journey*, presented by Dr. Carl Sagan. My high school physics teacher liked this documentary so much that he decided to let the class watch the entire series, a decision that earned him the approval of the class…that is, until he quizzed us on the material! Over the next several weeks we took in *Cosmos*. Though Carl Sagan was a well-known atheist, I did not feel from him a sense of aggression toward the believer, at least not as I watched the series. We lived in a small midwestern town, and it would be safe to say that most of the class had a basic understanding of Christianity and a fledgling faith.

As I watched *Cosmos*, something somewhat magical and ironic happened. The more I learned about the intricacies and subtleties of nature, the more my faith in God was strengthened. I don't think this was Dr. Sagan's intention. He described everything: humankind's experience of learning, discoveries from the ancient civilizations, Renaissance science, the great leaps of knowledge made in the twentieth century. The very small and the very big were illustrated: the first microscope with its discoveries, the subatomic level, even the extent of the known universe. Sagan described exceedingly large numbers. It was here I learned about the googol and googolplex,[1] and even the notion of infinity was explained. As I sat there in my high school classroom, I was not just learning; I was experiencing the universe in my mind.

But Sagan was also critical of religion. I remember his comments on how Roman Catholic Church leaders treated Galileo Galilei. It was the early seventeenth century and Galileo, a famous Italian physicist, proposed, among other things, that not all objects move around the Earth, and that the Earth moves around the Sun. This observation was heresy for the Church. On June 22, 1633, Church authorities handed down the following order:

> We pronounce, judge, and declare, that you, the said Galileo...have rendered yourself vehemently suspected by this Holy Office of heresy, that is, of having believed and held the doctrine (which is false and contrary to the Holy and Divine Scriptures) that the sun is the center of the world, and that it does not move from east to west, and that the earth does move, and is not the center of the world.[2]

I understood the juxtaposition here. *Cosmos* allowed me to experience the infinity of science on the one hand, and on the other hand, the small-mindedness of religion. I remember supporting Galileo in my mind, knowing that he did nothing wrong. I felt somewhat embarrassed for the Church. It was clear to me that Church leaders committed offenses as they misunderstood or poorly interpreted the Scriptures. Nowhere in the Bible is the Earth made to be the center of the universe. Galileo was right, the Church was wrong. The Bible had nothing to do with this entire conflict except that it was not read correctly.

Galileo was not a threat to the Church, nor to anyone else, for that matter. Given a chance, this seventeenth-century scientist was in a position to help people better understand the creation. But those doors were shut by religion. Despite my growing faith in God, I quickly took the side of science. My teacher and classmates felt the same. But something subtle was triggered, though I did not know what it was at the time. Fortunately for me, I didn't stop asking questions, not when it came to science or when it came to learning about God.

Thirty-five years later, the new version of *Cosmos*[3] appeared. Now there was a change in emphasis; the 2014 series *Cosmos* seized upon the widening rift between science and faith in our culture. Though this new series did not emphasize Church wrongdoings, there was considerably more emphasis on *naturalism*.[4] The presenter, Neil deGrasse Tyson, for whom I have considerable respect, made it clear that there was no room for God in the story of the universe. God or the gods of the past were used only to explain that which was unexplainable. The human experience has "matured," and we no longer need to live under the veil of superstition. Science fills the role church once took up. *Cosmos* also emphasized that the rift between science and faith is indeed a chasm. The host made it clear that one could traverse the chasm, but only in one direction: from the ignorance of faith to the enlightenment of science, so the faith-science gauntlet was again thrown down.

The leap

Science has transformed our lives. The application of scientific knowledge in the form of technology has turned our civilization into a wonderful place. Any earthling from any century before 1900 would undoubtedly be shocked if they could visit us today. But strides forward in technology and science are not always a smooth and continuous process. If we traveled back in time to the late seventeenth century in Cambridge, England, we would find one of the greatest scientists of all time, Isaac Newton. In his famous book, *Principia*, Newton described the motions of planets using his force equation. He also solidified the scientific concepts of a heliocentric solar system, a concept pursued earlier by Nicolaus Copernicus and Galileo Galilei. Newton's inverse law of gravity became the cornerstone of understanding planetary motion for two centuries. His law of gravity described how Earth moves around the Sun successfully; a better theory had replaced the earth-centered universe. In other words, Newton's theory replaced an approach established centuries earlier. Before Newton, science relied on gravitational ideas from the fourth-century-BCE Greek philosopher Aristotle, who stated that the motion of massive bodies tended toward the center of the universe or the downward direction.

Newton's concept of gravity was revolutionary. As the legend goes, he had a genius moment while sitting under an apple tree; an apple fell on his head. This stimulus led to the law of universal gravitation,[5] a single equation that described the force between masses and how planets and comets orbit the Sun.[6] Newton's formula later assisted astronomers in the discovery of Neptune. The universe was now opening up to humankind.

But as beautiful as Newton's formula ultimately proved to be, it was inaccurate. It was not until 1915 that Albert Einstein's new theory on general relativity required a small correction to the inverse square law acting on the two masses. Newton was on the right track, but he lacked the knowledge and the mathematical formalism that would be needed to improve his work.

What happened in 1915 to Newton's formula is the same as what happened when Newton improved upon Aristotle's concept of gravity. A refinement process is always at work in science, and it permeates every discipline. The hope is that the work of Aristotle, Newton, and Einstein will eventually lead scientists to an endpoint. Scientists dream that a single equation or set of equations will connect all forces, mass, time, and space.[7]

But this hope is tinged by the following reality: no matter how long the refinement process goes on, there is no assurance that the end will ever be reached. No matter how deep science probes, a question mark will eventually be found on the investigative path. We will never be able to get down to such a small dimension as to say our knowledge is now complete.[8] As we explore the vast cosmos, for example, we simultaneously reach back in time. But even if we could go back to the very beginning, our minds would entertain questions science cannot answer: What happened before that? Despite progress made in the past few centuries, there is a limit to what we can know. That limit is fast approaching.[9]

This realization may seem depressing, but there is an undeniably positive side to science: it never gives up. Science has a perpetual self-correcting mechanism that continues to allow us to move forward with discovery. If someone proposes an inaccurate theory, later on someone comes up with a better one. For this reason, I understand why many people put their trust in science.

Rings of knowledge

Science is based on knowledge, and that collective knowledge has been growing since the dawn of civilization. It can be helpful to imagine bundling all our acquired knowledge together to see where we are. We start with the sum of all possible knowledge, which we call the circle of knowledge. Unlike the circle of life, which rotates from birth to death, the circle of knowledge expands forever outward. The figure shows a simple sketch of all collective knowledge.

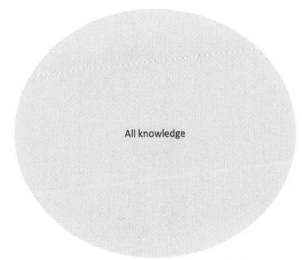

A circle containing all knowledge

Within the outer circle is every bit of knowledge that has ever been known or ever could be known. Within this circle is knowledge of the universe concerning both the living and the nonliving. Here we find all the intricate details about how humans work and think, and all understanding of the animal kingdom, the plant kingdom, and microorganisms. We find the knowledge of how life began, and all abstract concepts of all the mathematical and physical sciences. The circle also contains information about the smallest and largest scales of the physical universe, as well as knowledge of the most remote past and distant future.

Now inscribe in this circle another, smaller circle that represents knowledge to which we currently have access. In principle, this circle is easier to draw: in it we categorize all information presently available to us. It's reasonable to ask, how much smaller is this inscribed circle than the larger one? The answer to that is: we don't know. There will come a time when better theories replace tested ones, and as we have seen with the understanding of gravity, better ideas will replace old ones. So it's also reasonable to ask, will this process go on forever? *Epistemology,* or the study of knowledge, pursues questions like this, and the answer is: yes. Science is an ever-refining process toward the

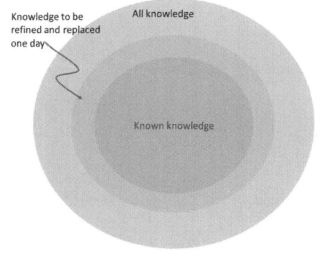

The circle of all knowledge with a smaller circle containing all current knowledge and a middle ring representing acquired knowledge that will eventually be corrected or updated

truth, and as science progresses, it's hard to say what is or isn't solidly correct. At least some, if not all, knowledge will be replaced by new knowledge as science progresses.

One might also imagine an inner ring representing knowledge that is currently understood to be correct. The difference here is that some knowledge will eventually be replaced by more up-to-date information. The natural refinement process of science will find our current knowledge either inaccurate, insufficient, or just wrong. Perhaps the most disturbing thing (or the most delightful thing, depending on your point of view) is that this intermediate annulus may be the entire inner circle.

The rings of knowledge are a humbling reminder that we are limited in our understanding and will forever be so. Though scientists continue to make fundamental discoveries, fundamental questions will stubbornly remain: Why are we here? What is our purpose? And yes: What is the meaning of life? Perhaps the most sobering aspect is this: science alone cannot answer the most fundamental questions.

The step

Contrary to the common understanding, people gain faith in steps, not leaps. And contrary to common opinion, faith has many similarities to science. Science starts by posing questions; then, we invoke our senses to make measurements. But there are limits to what we can measure, even with the help of instrumentation.

Likewise, understanding God is through our senses, not our faith, at least in the first step. The Bible spends no time trying to convince us that God exists. There is an unmentioned assumption that the universe has a Creator; evidence of this, according to the Bible, is ample. The book of Romans reminds us:

> Ever since the creation of the world his eternal power and divine nature, invisible though they are, have been understood and seen through the things he has made. So they are without excuse. (Romans 1:20)

Many Bible skeptics find this hard to accept: nature reveals God, and God reveals himself in nature. Of course, being labeled "without excuse" is not pleasant, so if you find yourself struggling with this statement from Romans, you are not alone.

The Bible also shows the human spirit: we need proof before we believe something. The Apostle Thomas demonstrates this. He was absent when Jesus revealed himself to his disciples following his resurrection:

> But Thomas (who was called the Twin), one of the twelve, was not with them when Jesus came. So the other disciples told him, "We have seen the Lord." But he said to them, "Unless I see the mark of the nails in his hands, and

> put my finger in the mark of the nails and my hand in his side, I will not believe."
> A week later his disciples were again in the house, and Thomas was with them. Although the doors were shut, Jesus came and stood among them and said, "Peace be with you." Then he said to Thomas, "Put your finger here and see my hands. Reach out your hand and put it in my side. Do not doubt but believe." Thomas answered him, "My Lord and my God!" Jesus said to him, "Have you believed because you have seen me? Blessed are those who have not seen and yet have come to believe." (John 20:24–29)

Whether you are a Bible believer or not, can you relate to Thomas? I certainly can: I will doubt unless I see the evidence. If someone claims to have resurrected from the dead, I want to see the dead body and then later see that person walking around in good health. Thomas, being a healthy skeptic, needed to see more evidence before he could believe this fantastic event had occurred. But take note that Jesus did not discipline Thomas because he needed evidence; he simply said people are extremely happy (blessed) if they can get past that point and believe without seeing *all* the evidence.[10] Thomas was not without any evidence before he saw Jesus in front of him. The other disciples had seen Jesus and enthusiastically believed, so you might expect that Thomas was at least optimistically hopeful that Jesus had come back to life, but seeing the dead rise doesn't happen that often.[11]

Having faith in something should perhaps be as natural as believing a scientific idea; both concepts rely on incomplete information about the world around us. It's up to us what we decide to believe and what we don't. So the real question is not whether we have faith; instead, what do we have confidence in and just how much evidence do we need?

Though Jesus said that we are blessed when we believe without seeing, he did not withhold evidence of his resurrection from Thomas. It took a week for Thomas' knowledge to catch up with the other disciples. When the famous encounter with Jesus finally did occur, the Teacher did not hesitate to show Thomas his hands, his side, and his living body. If there were any correction at all given to Thomas, it was that he had sufficient evidence to believe, yet did not. But none of us should look down on him. When he finally did come to the conviction that Jesus rose from the dead, his life was forever transformed.[12]

The miraculous year and the miracle life

Faith is not exclusive to Bible stories; it goes deep in science. Arguably, the most remarkable single year for modern physics was 1905, a year physicists refer to as an *Annus mirabilis*.[13] Though it happened to be the year my grandfather was born in Hamburg, Germany, the more newsworthy event occurred several hundred miles south, in Bern, Switzerland. This was the year that the young Albert Einstein, an unknown patent clerk at the time, wrote four papers in his spare time. Those four papers changed our way of thinking about the universe, space, time, mass, and energy. One of those papers was on special relativity.[14] Einstein postulated that the speed of light is constant in a vacuum and that nothing can travel faster than light. That same year Einstein also postulated that energy could be transformed into matter and matter into energy, $E = mc^2$. He also published a foundational paper on the effects of quantum mechanics that would eventually win him the Nobel Prize in 1921. In a fourth paper, Einstein wrote about the gravitational field that would open a path to general relativity (1915). This thread of research led to our modern understanding of the universe and completely shook the scientific world. Though Einstein's postulates were remarkable, many prominent scientists were not convinced.

Science is not complete until an idea can be proven. Most of what Einstein did in 1905 and his later work in 1915 had to wait years before it could be solidly verified. The first validation of general relativity came in May 1919 during a total eclipse of the sun. A star's position behind the sun was shifted as a result of light bending around the sun, a prediction of general relativity: the theory was validated! Relativity replaced the incorrect notion that light must travel through space as sound waves travel through the air. Following relativity, there was no longer a need for an invisible medium called *ether*.[15] Since that day, general relativity has stood the test of time and continues to be confirmed, though experimental evidence, including the prediction of gravitational waves, had to wait 100 years.[16]

More than a century after Einstein's busy summer of 1905, we view him as the physics guru of all time; his name has become almost synonymous with genius. Nonetheless, his ideas did not enter the mainstream thinking of the physics community without tremendous resistance.

Einstein did have some evidence to support his theory when he devised it, but there was no overall convincing experiment to prove it. Some prominent physicists, many from Britain, held fast to the nineteenth-century concept of ether. Stanley Goldberg, a physicist and historian, reminds us that the British physicists at the time had a "theoretical commitment to the ether." Even Lord Kelvin, who helped unify the emerging fields of physics in their modern form, argued in 1907 that ether must be an "elastic, compressible, non-gravitational solid."

Lord Kelvin and nearly the entire physics community were in error. Light always travels through a vacuum at a speed of 299,792.458 km/sec (or approximately 186,000 mi/sec). Other ideas from Einstein were also validated over time, not the least of which was the concept that energy and mass are interchangeable. The detonation of the first nuclear bombs clearly illustrated this.

How about faith? Is there anything we can learn from the leap of science brought about by Einstein? Can the advances of science made in the twentieth century help our discussion of faith? It's hard to usurp ideas, especially when people have invested considerable time in an idea that supports the status quo.[17] But we shall see in the upcoming chapters that nature has supplied evidence that allows one to take steps of faith, and we need not wait for years and years for data to come rolling in. Of course, what would impress people most would be tangible evidence of God himself. If we could only see him walking, breathing, touching, and feeling, would not our faith be propelled forward?

Jesus' life was a miracle. His enemies could not find traction in anything he did wrong. The only heresy they found to pin on him was his claim that he was God (e.g., Luke 22:70–71); this claim was his condemnation.[18] Religious leaders were so enraged by jealousy, they ended up bringing him before the Roman authority to have him crucified. Why Jesus, being all-powerful God in the flesh, would allow himself to be executed by mere humans is in itself a fascinating concept (see Part III).

The story of Jesus' life, death, and resurrection lends evidence for the scientifically curious. The appealing aspect of this story is that we don't need much faith to examine it. Historical records can be studied. The background accounts of Jesus' era, including concurrent kingdoms and civilizations, have all undergone rigorous archaeological scrutiny and scholarship.[19] Furthermore, the authentication of the Bible is also a well-studied topic. A leap of progress was made in this area over the twentieth century, primarily through the discoveries of the Dead Sea Scrolls. Among these scrolls was found the book of Isaiah, and in that book are references to events that occurred in Jesus' life. The Dead Sea Scrolls are also dated a few hundred years before Jesus walked the Earth. If evidence is what the scientifically curious mind needs to confirm God, then good news awaits.[20]

Chapter summary and questions
Some challenge the concept of faith in the scientific community. But there are limits to our knowledge—and limits to what science can achieve. Science depends on evidence and so does seeking God. Fortunately for the God-seeker, *the evidence is available to us today.* All of us can test the claims of faith in the God of the Bible.[21]

Questions to ponder:
- How can scientific facts challenge one's faith?
- Have you watched the two versions of *Cosmos* (the 1980 version and the 2014 version)? If yes, have you noticed a difference in tone toward the believer? Why do you think this is?
- Can you give one or two arguments that God does not exist? Could you explain your reasoning?
- Do you know of examples where fundamental science is reaching a limit? (In other words, do you know in which areas scientific progress is slowing more and more each year?)
- If you agree that there are limits to what science can do, how does this change the way you view faith?

NOTES

[1] The googol is 10^{100} or 1 followed by 100 zeros, a colossal number, far larger than almost anything imaginable in the universe. In comparison, the total number of elementary particles in the entire known universe has been estimated at 3.28 x 10^{80}. J. Bennett, "How Many Particles Are in the Observable Universe?" *Popular Mechanics*, 2017. But a googolplex is vastly larger than a googol. It can be conveniently written by stacking exponents $(10^{10})^{100}$, but if written out longhand, it is a number so large that using a 9-point font, it would not fit into the observable universe.

[2] Sentence of the Tribunal of the Supreme Inquisition against Galileo Galilei, given June 22, 1633. Public domain.

[3] *Cosmos: A Spacetime Odyssey*, director Brannon Braga, 20th Television, 2014.

[4] Naturalism: A philosophical viewpoint according to which everything arises from natural properties and causes, and supernatural or spiritual explanations are excluded or discounted.

[5] According to Newton, the force between two masses decreases with the inverse distance between the two. The force was also proportional to the product of the two masses. For example, using the metric system, the force is written as $F = Gm_1m_2/r^2$, where m_1 and m_2 are the masses of the two objects in units of kilograms, r the distance between the two objects in units of meters, and G is the universal gravitational constant, 6.67×10^{-11} newtons $(m/kg)^2$.

[6] Newton's formula was used to assist Edmund Haley (1656–1742) in calculating that the famous comet, previously observed in 1682, would return in 76 years. The comet did return as predicted in 1758, long after both Newton and Haley were dead.

[7] The hypothetical Theory of Everything (TOE) will combine all the aspects of the universe, including its four forces, strong, weak, electromagnetic, and gravity, into one single coherent theory.

[8] Quantum mechanics teaches us that there is a "fundamental" limit to a size called the Planck length or 1.6×10^{-35} meters. Physical reality becomes blurred or meaningless below this limit.

[9] Progress in fundamental science is ultimately constrained by what we can observe. Imagine approaching a wall (representing the limit to what we can know). On the other side of that wall is the unknown, that which can never be known. Attached to that wall are powerful springs (imagine a very stiff mattress). As we approach the wall, the springs compress and push back. The closer to the wall, the more the springs compress and resist. We will never actually arrive at the wall because the force of those springs will be overwhelming.

How does this work? Two examples come to mind; one concerns the very large, the other, the very small.

First, the very large. We can only observe a small part of the universe and three facts (springs) tell us the following:

- We live in a universe that has finite time. Though the universe is very old (13.8 billion years old) this is still a finite amount of time.
- The speed of light is finite. Though the speed is very large (186,000 miles/second or 300,000 kilometers/second), this value is still finite.
- Space and time (sometimes referred to as a space-time fabric) have expanded and stretched ever since the beginning of time (the Big Bang). This stretching causes distortions in what we observe far away. As we approach these boundaries the stretching of space and time distorts everything more and more the farther we look.

Together, these three conditions don't permit an observer access to any place in the universe; one is constrained.

There are also limits to what we can observe in the very small—another wall with its own stiff mattress awaits us there. Exploring the very small requires using more energy and propelling particles faster and faster before they collide in accelerators. The largest and most powerful accelerator in the world, the Large Hadron Collider (LHC) in Geneva, Switzerland, can accelerate protons to speeds of 99.999999 % of the speed of light. The LHC is 27 kilometers in circumference and probes to about 10 to 19 meters. But a particle accelerator powerful enough to probe the Planck scale (10 to 35 meters), a scale where physical meaning breaks down, would require an accelerator with a circumference similar in size to the orbit of Mars and constructed from about as much material as our moon. We are truly limited in what we can observe, both in the very large and the very small.

[10] On the other hand, those "full of faith" should be careful not to drive away others who need more time for their faith to materialize as they gather evidence.

[11] There have been few events like this in the previous 2000 years of Israel's history. The Bible records 10 separate occasions of the dead rising, including the son of the widow in Zarephath (1 Kings 17:17–24), the son of a woman in Shunem (2 Kings 4:18–37), an Israelite man (2 Kings 13:20–21), the son of the widow from Nain (Luke 7:11–17), Jairus' daughter (Luke 8:49–56), Lazarus (John 11:1–44), saints in Jerusalem (Matthew 27:50–54), Tabitha (Acts 9:36–42), Eutychus (Acts 20:7–12), and of course, Jesus Christ himself (Matthew 28:1–20; Mark 16:1–20; Luke 24:1–49; John 20:1–21:25).

[12] Legend reports that Thomas traveled by boat to India, where he preached that Jesus rose from death. His message led to his martyrdom.

[13] Latin for "wonderful year."

[14] Albert Einstein, "Zur Elektrodynamik bewegter Körper," *Annalen der Physik* 17:891, 1905. English translation: On the Electrodynamics of Moving Bodies.

[15] Before special relativity, a mysterious substance called ether was thought to fill all space. Ether formed a medium for light to travel through, much like sound waves go through air, or ocean waves move through water.

[16] Gravity waves were experimentally found in 2017 in the LIGO (Large Interferometer Gravity Observatory), where this author performed a small task, building and testing a stabilizing mechanism for the prototype apparatus in 1985.

[17] There is a parallel with science here: evidence ultimately made Einstein world famous. Without evidence, he would have been known as just another good physicist with interesting theories. The physical measurement, starting with

observing how light from a distant star "bent" around the sun during a solar eclipse in 1919, provided considerable evidence that relativity was real. Einstein later received the Nobel Prize in 1921 (which incidentally was on another idea, quantum mechanics, not relativity).

[18] See Mark 14:55, 7:37, 14:57–59; Luke 19:47–48, 23:15.
[19] More on these topics in Chapter 8.
[20] More on this topic in Appendix 4: "A Proof of Chronology: The Dead Sea Scrolls."
[21] More on faith in Appendix 1: "Faith: biblical and scientific – two sides of the same coin."

Chapter 3 – Tracing Earlier Steps

You cannot connect the dots looking forward; you can only connect them looking backward. – Steve Jobs

Humans ask questions; it's in our nature and always has been. Long before iPhones, iPads, and the internet came to be, we occupied our minds with things around us. And if you wander outside on a clear moonless night far from a city, lie down on the grass, and look up to the sky, you might just notice something you never have seen or perhaps not seen in a while: the beauty of the pristine heavens, stars with color, planets, constellations, one of the visible nebulae, maybe even a shooting star. Imagine if we could only leave our busy world, just for an evening, and go back to a past when everything moved a bit slower. No, I'm not talking about going back millions of years, when men lived in caves. I am thinking about going just far enough back to when society had more opportunity to reflect on the world, reflect on our origins, and reflect on God. I am thinking about a state of mind and a place without the constant interference from noise and light that fills our eyes and ears day and night. For a scientist who sought God, this was a special time.

A few years ago, I had the good fortune of finding myself in a state of mind like this. I was on a mathematics research trip in Adelaide, Australia. I had arrived a few days earlier and had not yet recovered from jet lag—my mind and biorhythms were 12 hours offset. For many years I have had the habit of waking up early each morning to go out for a run, and that morning I woke up exceptionally early, somewhere around 3 am. I headed outdoors and did not need to go far from the city center to find a park that was dark enough to observe the stars in the very early morning. There was a brilliant display of "new" stars, ones that I had never seen with my own eyes.

I slowly panned my head to get a good look at each part of the sky. On the flat horizon were the eucalyptus trees on the edge of the field, and as I looked up, I could see the Southern Cross, a constellation marking out a crucifix and pointing toward the south celestial pole. The full band of the Milky Way Galaxy stretched across the entire sky. The first light of the sun had to wait another few hours before it would come over Adelaide's eastern hills. The city was asleep; I heard the odd chirping of native birds and I stood for a moment to breathe in the fragment aroma of the trees. I felt almost as if I could interact with my surroundings—the feeling was surreal. It was easy to contemplate thoughts of eternity as I closed my eyes.

The Milky Way Galaxy as seen from South Australia (Source: Australian Public Broadcasting)

Moments like that morning in the park in Adelaide don't occur nearly often enough, at least not for me. But if we can somehow turn down the noise, we might be able to hear nature whispering to us. Scientists of the past had this luxury, and I envy them.

A few centuries ago, the latest news would come by town crier, and entertainment would come from books, the theater, or better yet, only from nature. For the scientist, this was an enchanting time to live;

you didn't need to be awarded millions of dollars in grant money or be a part of a sizable collaboration of scientists to advance fundamental science.

As I studied great scientists of past centuries, I discovered that many were men of faith. Their views connected God to their understanding of the world around them, and this was not because they didn't have an explanation as to how things worked (God of the gaps again); rather, many of them had sophisticated ways of viewing nature. Faith was as much a part of the lives of these scientists as was their pursuing deep questions about the natural world.

Nicolaus Copernicus

Nicolaus Copernicus (1473–1543) was one such scientist. This Renaissance polymath, with wide-ranging knowledge and learning, helped shape our understanding of our world and aided us in rethinking our place in the universe. He eventually developed a model of the planets as they went around the Sun and put the Sun, not the Earth, in the center of the solar system. Copernicus also believed in God, though he was part of a church that struggled deeply with its practice. This church had perpetrated two centuries of atrocities during the Crusades and was still practicing an extreme form of religious intolerance and repression through the Spanish Inquisition. Among the least of the church's issues was its incorrect interpretation of the biblical text. Religious leaders used poetic verses in the book of Psalms to support nonphysical versions of nature. They viewed the Earth as an unmovable object, and therefore, all other heavenly bodies must be circling it.[1]

Despite the significant advances in the Renaissance, few people were educated; only about one in ten could read.[2] The lack of education meant that a tiny minority could ask critical questions that challenged the contemporary understanding of nature or existing interpretations of the Bible. The observations that Copernicus made on the rooftop of his home were not extraordinarily complicated. The science that he performed could have handily been done centuries earlier by someone with curiosity. The young Polish astronomer used simple instrumentation as he plotted the planets changing their position, night after night, relative to the stars. Yet the discoveries of Copernicus were astounding and completely new. It was his broad knowledge and understanding of Scripture that prevented his astronomy work from being misinterpreted.

The remarkable discoveries by this Polish scientist came only after difficulty in his early life. Most of us did not pursue the academic track that Nicolaus did, but I am sure we can relate to his starts and stops as well as his ups and downs in life. He lost his father when he was only 10 and his mother a decade later. His maternal uncle, Lucas Watzenrode (the Younger), a wealthy and influential merchant, took the boy under his wing and saw to his education and career.[3] All of a sudden, the opportunity to study arithmetic, geometry, geometric optics, cosmography, and theoretical and computational astronomy were all available to the young man at the University of Kraków. Nicolaus had exposure to the Greek philosophers, including the philosophical and natural science work of Aristotle[4] and Ptolemy,[5] and the Muslim Andalusian astronomer Ibn Rushd.[6] This broad academic exposure stimulated his interests. Eventually, the young astronomer was able to rethink the models of Ptolemy that had stood the test of time for fifteen centuries, replacing them with a more straightforward heliocentric solar system where the Sun, not the Earth, was in the center.

This revelation of Copernicus did not happen in an instant. His model materialized over time. It's easy to imagine that he cycled between the Bible, academic books, and outdoor observation, all the while asking many questions. After his early education in Poland, his uncle's search for power helped Nicolas even further, though perhaps unwittingly. In those days, a position with the Church was associated with power. Lucas Watzenrode was politically savvy and understood the governing system. After he was elevated to Prince-Bishop of Warmia (northern Poland), he tried to position his nephew in the Warmian canonry, a move that would help Watzenrode secure his power.[7] As fate would have it, Watzenrode was opposed, delaying Nicolaus from assuming this comfortable Church position. But Watzenrode used this time wisely and had Nicolas sent to learn canon law in Bologna, Italy.

The young, scientifically minded man went off to college, and his interests changed.

During his three-year stay at Bologna, between fall 1496 and spring 1501, Copernicus seem[ed] to have devoted himself less keenly to studying canon law...than to studying the humanities—probably attending lectures by Filippo Beroaldo, Antonio Urceo, called Codro, Giovanni Garzoni, and Alessandro Achillini- and to studying astronomy.[8]

On his second of two sojourns in Italy, Copernicus eventually received his degree of Doctor of Canon Law in 1503. Historians Dobrzycki and Hajdukiewicz tell us that this was not before ideals crystallized in his mind of a "new view of the heavens." When he returned to his home in Warmia, he worked as a secretary for his uncle, but also made measurements of the position of the planets. "Copernicus observed the five visible planets, Mercury, Mars, Venus, Jupiter, and Saturn, all within a fraction of a degree of accuracy, all without a telescope."[9] In his position as administrator, Nicolaus continued with his scientific research. By 1514, he published his heliocentric model. He did this while he continued with all his ecclesiastic and administrative duties for the Church and the state.

Eventually, Copernicus presented his model to the highest level in the Church, to Pope Clement VII in 1533,

...who approved, and urged Copernicus to publish it around this time. Interestingly, the scientist was never under any threat of religious persecution and was urged by both Catholic and Protestant leaders. Copernicus sometimes referred to God in his works, and did not see his system as in conflict with the Bible.[10]

One might ask: How did Copernicus reconcile his scientific discoveries with his belief in God? The simple answer is that Copernicus probably understood the Bible well enough; most likely, he read it for himself.[11]

Unfortunately, most of the population did not have access to the education Nicolaus Copernicus had. The average layperson in the sixteenth century had to rely on the clergy and their interpretations of the Bible. This situation was complicated as the Church struggled more and more to retain power. As the Protestant Reformation gained traction in Europe, power shifted away from Rome. One way to hold power was to ascribe authority through rigid interpretations of the Scriptures. Copernicus just missed the harshest backlash from the Church, but scientists who followed his example were not so lucky.

Astronomer Copernicus, Conversations with God, by Matejko

Galileo Galilei, Isaac Newton, and many others

One of the unluckier scientists from this era was Galileo Galilei (1564–1642). Galileo is considered one of the giants of astronomy and the father of observational astronomy; he made the first telescope observations of Jupiter's moons (1610), and his discovery supported the underlying assumptions of Copernicus. Galileo observed that some objects orbit other planets, debunking the Earth-centered universe. Again, Galileo found no contradictions between his observations and the Bible. He was known to have said, "The Bible cannot err," and he saw his system of planetary behavior as an alternate interpretation of the biblical texts.[12] Unfortunately, it took centuries for the Church to see its wrong position on science. Galileo was eventually vindicated by the Church long after his death.[13]

As science advanced, more discoveries came, and many went against the Church's interpretation of the Bible. Meanwhile, many scientists well versed in both science and biblical knowledge wrote clear statements on their faith as they pushed forward with astounding discoveries. Sir Francis Bacon (1561–1627), a philosopher, is known for establishing the scientific method of inquiry based on experimentation and inductive reasoning. In his work, *The Interpretation of Nature*,[14] Bacon established his goals as discovering the truth, service to his country, and service to the church. A devout Anglican, he rejected atheism as being the result of insufficient depth of philosophy. He expressed his conviction of creation from God in his *Essays*:

> God never wrought miracle, to convince atheism, because his ordinary works convince it. It is true, that a little philosophy inclineth man's mind to atheism, but depth in philosophy, bringeth men's minds about to religion.[15]

Bacon is also known for his famous quote in which he addresses science and faith together:

> To conclude, therefore, let no man...think or maintain, that a man can search too far, or be too well studied in the book of God's word, or in the book of God's works; divinity or philosophy; but rather let men endeavor an endless progress or proficience in both.[16]

In the same period, other renowned scientists made revolutionary discoveries and continued to express their sincere spiritual convictions. Johannes Kepler (1571–1630), a brilliant mathematician and astronomer, established the laws of planetary motion about the Sun. He wrote,

> We see how God, like a human architect, approached the founding of the world according to order and rule and measured everything in such a manner.[17]

A contemporary of Kepler, René Descartes (1596–1650) was a French mathematician, scientist, and philosopher. Descartes saw the existence of God as central to his whole philosophy. At the age of 24, Descartes had a dream that inspired him to write, "I think, therefore I am"; yet, in his philosophy, that famous quote could not be upheld if God was not a certainty.[18]

Many scientists consider Isaac Newton (1642–1727) to be the most celebrated physicist and mathematician before the twentieth century. In Newton's system of physics, God was essential to the nature and absoluteness of space. In *Principia,* he stated, "The most beautiful system of the Sun, planets, and comets, could only proceed from the counsel and dominion of an intelligent and powerful Being."[19]

James Clerk Maxwell (1831–1879) was perhaps the most well-known physicist of the nineteenth century. He developed some of the most fundamental equations of electricity and magnetism; he also had a deep respect for God and an extensive knowledge of the Scriptures from his childhood. Maxwell knew the chapter and verse of almost any quotation from the Psalms (by age eight, he had memorized Psalm 119). He visited the sick and prayed daily with his wife and family, and in his letters to his wife, he discussed Scripture passages. Maxwell was also convinced of the compatibility of the Bible with scientific investigation, brought out in this prayer that he wrote:

> Almighty God, Who hast created man in Thine own image, and made him a living soul that he might seek after Thee, and have dominion over Thy creatures, teach us to study the works of Thy hands, that we may subdue the earth to our use, and strengthen the reason for Thy service; so to receive Thy blessed Word, that we may believe on Him Whom Thou hast sent, to give us the knowledge of salvation and the remission of our sins. All of which we ask in the name of the same Jesus Christ, our Lord.[20]

There have been many other scientist-believers over the centuries; their stories, discoveries, and solid faith in God could fill the pages of another book.[21]

Albert Einstein

One scientist changed the way we viewed the heavens and even our own reality—that person was Albert Einstein (1879–1955). Einstein is considered by many to be among the greatest, if not the greatest physicist of all time, and he continues to be revered for his revolutionary ideas about time, gravity, and the conversion of matter to energy. His famous equation $E = mc^2$ represents Einstein in the minds of many, though this is only a tiny fraction of his life's work.[22] Though Einstein probably never believed in a personal God,[23] we should take notice of his thoughts and interests, and in particular, what hindered him from going further in his faith.

In 1936, a sixth-grade girl named Phyllis wrote him a letter on behalf of her Sunday School class. "We have brought up the question," she wrote, "Do scientists pray?"[24]

Here is Einstein's reply:

January 24, 1936

Dear Phyllis,

I will attempt to reply to your question as simply as I can. Here is my answer:

Scientists believe that every occurrence, including the affairs of human beings, is due to the laws of nature. Therefore a scientist cannot be inclined to believe that the course of events can be influenced by prayer, that is, by a supernaturally manifested wish.

However, we must concede that our actual knowledge of these forces is imperfect, so that in the end the belief in the existence of a final, ultimate spirit rests on a kind of faith. Such belief remains widespread even with the current achievements in science.

But also, everyone who is seriously involved in the pursuit of science becomes convinced that some spirit is manifest in the laws of the universe, one that is vastly superior to that of man. In this way the pursuit of science leads to a religious feeling of a special sort, which is surely quite different from the religiosity of someone more naïve.

With cordial greetings,
your A. Einstein[25]

In his letter to Phyllis, Einstein showed that he recognized man's limits to understanding and that there is a relationship between faith and science. Though he did not believe in the God of the Bible, he was not an atheist. Instead, Einstein believed in an impersonal God, one that brings harmony in nature, a belief following the ideas of the Dutch philosopher Spinoza.[26]

> I believe in Spinoza's God who reveals himself in the orderly harmony of what exists, not in a God who concerns himself with fates and actions of human beings.[27]

Einstein recognized the connection between faith and science, and he also realized that some things could not be explained. He phrased it as "a spirit manifest in the laws of the universe." This "spiritual insight" coming from this scientist may seem unusual, but is it?[28]

As a young man, Einstein was exposed not only to Judaism through his secular Jewish parents, but he was also given Christian instruction from the local Catholic public elementary school in Munich.[29] In his *Autobiographical Notes*, Einstein wrote that he was deeply religious until the age of 12, which then "came to an abrupt end."[30] From that time forward, he regarded the existence of an anthropomorphic God with skepticism and described the stories of religion as "naïve"[31] and "childlike."[32] Perhaps Einstein's skepticism of religion grew out of the condition of the world around him. He witnessed the long-lasting corruption between the state and religion, and this certainly accelerated his departure from all things religious. He commented that he reached:

> ...the impression that youth is intentionally being deceived by the state through lies; it was a crushing impression. Mistrust of every kind of authority grew out of this experience, a skeptical attitude toward the convictions that were alive in any specific social environment—an attitude that has never again left me, even though, later on, it has been tempered by a better insight into the causal connections.[33]

Religion might have left Einstein early on, but not his desire to understand God. One of his most famous quotes underlines his desire:

> I want to know how God created this world. I am not interested in this or that phenomenon, in the spectrum of this or that element. I want to know his thoughts; the rest are details.[34]

Albert Einstein on his sailing boat, 1936 (Source: The American Physics Institute)

As Einstein progressed in his work, his theories redefined the entire framework of our thinking about everything physical, space, time, mass, and energy. His ideas inadvertently helped him recognize the impossibility of a noncreated world: his general theory of relativity meant that the universe must have had a beginning.[35] This view clashed with the atheistic position at the time that the universe was infinitely old (avoiding the unsettling reality of a creator). Einstein forced a giant leap forward in the scientific community. Interestingly, this "new" view of nature as having a beginning perfectly rejoined the creation concept as described in the book of Genesis (see Chapter 6).

Einstein's early religious education probably meant that he wrestled with the concept of God throughout his life. He must have been familiar with the creation account written some 38 centuries earlier, probably by Moses. Einstein even had exposure to the stories of Jesus, and he undeniably accepted his historical existence. The great scientist responded to a question posed to him in 1929, "You accept the historical existence of Jesus?"

> "Unquestionably! No one can read the Gospels without feeling the actual presence of Jesus. His personality pulsates in every word. No myth is filled with such life."[36]

With all the thought and patience Einstein must have put into his contemplation of God, his view of God remained only that:

> My views are near those of [pantheist] Spinoza: admiration for the beauty of and belief in the logical simplicity of the order which we can grasp humbly and only imperfectly. I believe that we have to content ourselves with our imperfect knowledge and understanding and treat values and moral obligations as a purely human problem—the most important of all human problems.[37]

This profound philosophical valley, described by Einstein, is the place where many well-intentioned people, scientists included, get stuck. For Einstein, the complexity of the universe, its subtleties and order, implied that if there was a "spirit" that oversaw it, the universe had to be intrinsically good, or as he put it, "a universe which is both good morally and perfect physically." But that is not the universe we live in: evil most certainly exists, and what Einstein could not accept was that God would design the universe to allow evil to exist without some purpose.

Despite Einstein's misunderstanding about evil, what I appreciate about him, apart from his tenacity to rethink the physical world, is that he was initially on the right track about God (or what he referred to as a spirit). Einstein lived in a time and place that was fertile for creative ideas. Religious institutions were not so powerful as in previous centuries. Einstein had total freedom to consider new ideas, and, at the same time, he had access to the nineteenth-century advances in the sciences. He also had enough spiritual material available to him and could carefully work through the concept of God on his own without being told what to think.

Unfortunately, what Einstein found at the end of his logical journey is the same thing many sincere scientists find at the end of their search for God: an open question that leads to agnosticism.[38] This is the end of the journey for many scientists and nonscientists alike, and many religious and nonreligious people as well. Even many of those who claim to be atheists, when pressured, turn out to be agnostics, those who cannot say what is known or cannot be known about God. I do not look down on Einstein or anyone else for coming to this conclusion about God. Without understanding the nature of God, I too found it nearly impossible to get past that looming question mark at the end of the agnostic path.

We can benefit from Einstein's careful reflections, but what he failed to grasp was not the "what," it was the "who." According to Christianity, God is behind the scenes giving purpose to the entire universe and desiring a relationship with his creation: a relationship based on love. A relationship of this kind only exists with free choice; and free choices can lead to evil. Only a tiny fraction of creation is advanced enough to make a rational decision to choose between good and evil; most of the creation is untouched by flawed choice and remains in a pristine state with all its complexity, subtleties, and order. The fascination with the physical universe, unflawed by evil, is where Einstein stopped—and this is where many of us stop.

But many of us have yet to begin to take those first steps that Einstein took. One way to take the first step is to open one of God's books. We'll start with the book that is available to us as soon as we open the door and look outside, "the book of nature." This is where we'll head now.

Chapter summary and questions

Many of the greatest scientists of the past believed in God, not as a crutch or out of laziness, but by exerting effort as they studied the Scriptures. They did this while making significant scientific progress. Some scientists proposed ideas that may have been different from those held by church leadership, but those ideas did not differ from what the Bible teaches.

Questions to ponder:
- When was the last time you took a day off and just did "nothing"? How did it feel? Did anything creative come out of this experience?
- Please fill in the blank if this applies to you: I have not pursued seeking God because _____. If you identify what goes into the blank, how do you resolve to go beyond that barrier? If you believe this wall is insurmountable, why do you think that?
- Bonus question: Do some research on Spinoza's God. What are the characteristics of this God, and how do they differ from those of the God of the Bible? Could you elaborate on why Einstein might have been attracted to these characteristics of God and could not accept the God of the Bible?

NOTES

[1] For centuries, the Church was confident in its view of the Earth being the unmovable center of the universe. The sources for this position included: Joshua 10:12–13; 2 Kings 20:11; Psalm 93:1, 104:5; Ecclesiastes 1:5; Isaiah 30:26, 38:8; and Habakkuk 3:10–11. However, none of these scriptures support the view that the Earth is the center of the universe nor that it is physically unmovable. Church leaders did not consider the context. The poetic description of our planet, for example, "He set the earth on its foundations; it can never be moved" (Psalm 104:5 NIV), no more implies that the Earth is motionless than do other passages in the Bible suggest that God has wings like a bird (e.g., in Psalm 17:8, 36:7, 57:1). More importantly, the Bible was written with the worldview of the ancient Hebrews. More will be said on this in Chapter 7.

2 Max Roser and Esteban Ortiz-Ospina, "Literacy," published online at OurWorldInData.org, 2018, retrieved from https://ourworldindata.org/literacy in November 2018.
3 Jerzy Dobrzycki and Leszek Hajdukiewicz, "Kopernik, Mikołaj," *Polski słownik biograficzny* (Polish biographical dictionary), vol. XIV (Wrocław: Polish Academy of Science, 1969), 3–16.
4 Aristotle's two major works on astronomy are: "On the Heavens" (Latin: *De Caelo* or *De Caelo et Mundo*), and "Metaphysics," (Latin: *Metaphysica*), circa 350 BCE.
5 Claudius Ptolemy (100–168 CE), who established the concept of the (incorrect) epicycle model, a geometric construct used to explain the variations in speed and direction of the apparent motion of the moon, sun, and planets.
6 Ibn Rushd (1126–1198), Latinized as Averroes, an Andalusian philosopher and thinker who wrote about many subjects, including philosophy, theology, medicine, astronomy, physics, Islamic jurisprudence and law, and linguistics. Encyclopedia Britannica, https://www.britannica.com/biography/Averroes.
7 Dobrzycki and Hajdukiewicz, *Polski słownik biograficzny*, 5. The canonry, a prestigious nonclerical position subject to the ecclesiastic rule, would provide for a comfortable situation and a steady income and would increase the overall influence of the family.
8 Ibid.
9 Olaf Pedersen, "The Decline and Fall of the Theorica Planetarium: Renaissance Astronomy and the Art of Printing," in *Science and History: Studies in Honor of Edward Rowen* (Wroclaw: Polish Academy of Science, 1978), Studia Copernicana 16.
10 http://www.godandscience.org/apologetics/sciencefaith.html, last modified December 8, 2011, last accessed November 2018.
11 Copernicus was probably not disturbed by the Bible's geocentric view, and this may be connected to his understanding of the context in which the Bible was written (see Chapter 6).
12 Peter L. Gregory, *Laddie's Grave* (www.xulonpress.com 2010), 31.
13 More than 350 years after the Roman Catholic Church condemned Galileo, Pope John Paul II rectified one of the Church's most infamous wrongs: the persecution of the Italian astronomer and physicist for proving that the earth moves around the sun. Alan Cowell, "After 350 Years, Vatican Says Galileo Was Right: It Moves," 1992, New York Times (archive).
14 Francis Bacon, *De Interpretatione Naturae Prooemium*, compiled and published in 1653.
15 Francis Bacon, *The Essays*, 1597.
16 Francis Bacon, *The Advancement of Learning*, 1605. Copan and Jacoby also point out that Bacon warned, "We must not unwisely mingle or confound these learnings [science and faith] together." Paul Copan and Douglas Jacoby, *Origins: The Ancient Impact and Modern Implications of Genesis 1–11* (New York: Morgan James, 2019)
17 John Hudson Tiner, *Johannes Kepler: Giant of Faith and Science* (Fenton, MI: Mott Media, 1977), 178.

[18] René Descartes, summary, *Meditationes de Prima Philosophia* (*Meditations on First Philosophy*), in *qua Dei existentia et animæ immortalitas demonstratur*, 1641.
[19] Isaac Newton, *The Principia: Mathematical Principles of Natural Philosophy*, 1687.
[20] James Clerk Maxwell, in a prayer found among his notes, J.C. Maxwell, "Discourse on Molecules," a paper presented to the British Association at Bradford in 1873, as cited in E.L. Williams and G. Mulfinger, *Physical Science for Christian Schools* (Greenville, SC: Bob Jones University Press, 1974), 487.
[21] Some famous names among this list include: Blaise Pascal (1623–1662), Robert Boyle (1627–1991), Michael Faraday (1791–1867), Gregor Mendel (1822–1884), George Gabriel Stokes (1819–1903), and Max Planck (1858–1947).
[22] This famous equation is not even written out in its most complete form. The full form is $E^2 = m_o^2 c^4 + p^2 c^2$, but this reduces to the well-known expression when momentum is zero.
[23] Helen Dukas and Banesh Hoffmann (eds), *Albert Einstein, The Human Side* (Princeton University Press, 1981), 43.
[24] Letters of Note, "Dear Einstein, Do Scientists Pray?" Friday, May 18, 2012.
[25] Dukas and Hoffman, *Albert Einstein, The Human Side*, 32–33.
[26] Baruch Spinoza (1632–1677) was a Dutch philosopher and one of the early thinkers of the Enlightenment period. He came to be considered one of the great rationalists of seventeenth-century philosophy.
[27] Cablegram reply to Rabbi Herbert S. Goldstein's (Institutional Synagogue in New York) question to Einstein, "Do you believe in God?"
[28] Christians, of course, relate spiritual phenomena to the third person of the Trinity, the Spirit of God.
[29] Ralph Baierlein, *Newton to Einstein* (Cambridge: Cambridge University Press, 1992), 201–202.
[30] Albert Einstein, "Notes for an Autobiography." *Saturday Review of Literature* Volume XXXII, No. 48 (Nov. 26, 1949): 9.
[31] Alice Calaprice, *The Expanded Quotable Einstein* (Princeton University Press, 2000), 218.
[32] Michael R. Gilmore, "Einstein's God: Just What Did Einstein Believe About God?" *Skeptic* 5:2 (1997): 64, quoting a Sept. 28, 1949 letter to Guy Raner Jr., also July 2, 1945 letter to Guy Raner Jr.
[33] Einstein, "Notes for an Autobiography."
[34] Calaprice, *The Expanded Quotable Einstein*, 202.
[35] Einstein hesitated to believe the consequences of his own theory and incorrectly put in a constant term in his equations that would allow for an infinite universe (years later, he jokingly referred to this as his biggest blunder).
[36] George S. Viereck, "What Life Means to Einstein," *Saturday Evening Post*, 26 October 1929; Schlagschatten, Sechsundzwanzig Schicksalsfragen an Grosse der Zeit (Solothurn, Schweiz: Vogt-Schild, 1930), 60; and *Glimpses of the Great* (New York: Macauley, 1930), 373–374.
[37] Gerald J. Holton and Yehuda Elkana, *Albert Einstein: Historical and Cultural Perspectives* (Mineola, NY: Dover, 1997), 309.

[38] Agnosticism is the view that the existence of God, the divine, or the supernatural is unknown or unknowable.

Part II – A Leap in Science...

The second part of this book is about exploring the evidence for a Creator via the sciences—we'll look at physics, astronomy, and abiogenesis. We caution the reader that some of the material here is a bit technical. You may find the brief chapter summaries at the end of these chapters to be helpful, or you may even prefer to jump to Part III and then come back to Part II later. Otherwise, let's get started!

Chapter 4 – Our Fine-Tuned Universe

It is the glory of God to conceal things, but the glory of kings is to search things out. (Proverbs 25:2)

Our universe is a universe of many extremes, but at first glance, things around us seem to work intuitively: we push a box, it moves; we drop a pencil, it falls. The harder we push the box, the faster it moves, and the longer we watch the pencil fall, the faster we see it drop, but beneath the simple realities are complex subtleties of nature. Most of the fundamental realities of the universe are nonintuitive, and we can only observe them at a microscopic scale (fantastically smaller than the atom) or a colossal scale (considerably larger than our galaxy). Fascinating things happen at the middle scale too. The building blocks of the universe (molecules and atoms) and their subatomic particles (protons, neutrons, and electrons), effectively work together so that the ultimate creation in the universe, life, can exist. A cursory glance at our world could persuade some of us that nature seems to work well, elegantly, and beautifully. But if one digs deeper into the mechanics of things, they would be surprised and perhaps even shocked to see how matter all came together, often in very subtle ways, so that we can be here. But to see this, we need to first look at the big picture.

The cosmos is the entire physical universe, that which is visible and invisible to us, and it contains all that has ever existed. To begin, it helps to have a brief view of how nature itself is constructed so that we can truly appreciate the complexity of the cosmos. The basic building block of matter, the atom, is no larger than half a nanometer across, 5×10^{-10} meters. Atoms are mostly space, in fact, 99.9999999999996 % space. The atomic nucleus is composed of protons and neutrons, collectively called nucleons, each

53 | A leap in science

approximately a femtometer across, or 10^{-15} meters, each with about the same mass of about 10^{-27} kilograms. These small particles have few properties that grab our attention, but they are essential for almost everything we see. The proton, for example, has a positive charge, and the neutron has no charge; multiples of these two particles make up the nucleus of the atom. The electron, a substantially less massive particle than the proton (1836 times less), has an equal charge to that of the proton but opposite in sign (negative). Electrons occupy the space around the atomic nucleus, and powerful forces of nature keep them in regularly spaced orbits.[1]

The structure of the atom is also relatively simple, and this concept is repeated for every atom throughout the universe. The lightest atom, hydrogen, is composed of only a single proton, no neutrons, and a single electron that moves in a cloud around the proton. Only 92 different types of atoms occur in nature, and a few dozen more have been synthesized in the laboratory. Uranium is the most massive naturally occurring element, with 92 protons, 92 electrons, and 146 neutrons.

Now that we have been introduced to our nuclear particle friends, what can we ask about them? Why are there only 92 types of atoms and no more? What keeps the electron with a negative charge from rushing toward the proton with a positive charge? What holds the protons together in the nucleus, since particles with the same charge should repel each other? What is inside a proton or a neutron? Most of these questions currently have no answer.

Before the advent of modern physics in the early part of the twentieth century, physicists, chemists, and biologists made slow progress in understanding the world around them. However, with the emergence of modern science, large telescopes, accelerators, and electron microscopes, science has made great leaps forward. Along with these modern instruments, physical models have entirely revolutionized how we picture the world in its two extremes. On a small scale, the Standard Model of physics gives us a picture of the building blocks of protons and neutrons—quarks.[2] On a large scale, general relativity allows us to model the universe. In the last hundred years, science has created a clear picture that spans from the size of the proton to the known universe (a range of 60 orders of magnitude!).

As we shall see, the universe is made with incredible precision, so that through time, life—even intelligent life—can exist.

Forces in a fine-tuned universe

Just four forces guide everything physical: this applies for a proton, a neutron, an electron, an atom, a molecule, a rock, a planet, a star, or a galaxy. Most of us are familiar with two of these forces: gravity and electricity (or more appropriately, the electromagnetic or EM force). The other two forces, the strong and weak nuclear forces, were discovered only in the past century, and they dominate at the nuclear scale.

The strength of each fundamental force is seemingly "tuned" in a set of fundamental constants. The term "fined-tuned universe" describes the proposition that life in the universe can only occur when everything is just right. For a physicist, this means that the values of the fundamental constants are within a very narrow range. If they were outside of that range, the conditions would not be right for atoms to form, let alone molecules, rocks, planets, stars, galaxies, and, of course, life.[3] Just how precisely "adjusted" these values must be for us to be here presents one of the biggest mysteries in physics.

Gravity is the first force we meet. As infants, we attempt to stand for the first time, and gravity pulls us to the ground. Objects adhere to the ground and don't just float away. An object's mass exerts an attractive force field around it; small masses exert weak force fields, and larger masses exert stronger force fields.[4] An apple, for example, falls from a tree under the force of gravity. The apple's mass is attracted to the mass of the Earth, and they pull each other together. Since the mass of the apple is so much less than that of the Earth, the apple will move toward the Earth and speed up (accelerate) as it falls. The Earth, on the other hand, will rise toward the apple just a tiny bit. The difference in masses will mean that the apple does most of the moving.

All objects with mass are under the influence of gravity. The force that acts on the apple is the same force that causes the coalescence of space material (dust and gas) to form planets, our solar system, galaxies, and large-scale structures in our universe. The strength of gravity diminishes by the inverse square of the distance of the two masses acting on each other, or proportional to $1/r^2$, where r is the distance. The force of gravity, F_g, is expressed in an equation:

$$F_g = G \frac{m_1 m_2}{r^2}$$

—where m_1 and m_2 are the values of the two masses, and G is the universal gravitational constant.[5]

The inverse square law means that the force of gravity drops by a factor of four when the distance between the masses doubles. The force is again halved each time the distance continues to double, diminishing in strength by 1/16, then 1/64, then 1/256, as the separation is increased by 4, then 8, then 16, and so on. The small value of G implies that this force is feeble, and indeed, compared to the other three forces, it's exceptionally weak. Though the attractive force of gravity is weak and diminishes at a distance, it's nonetheless the most influential force on the largest scales. Planets, stars, galaxies, and ultimately, the universe, are all controlled at the large scale by gravity.

We are also familiar with the second of the four forces: the electromagnetic force. All of us know what happens when you rub a balloon on a sweater: the balloon sticks to the sweater. If you rub the balloon on your hair, assuming you have hair, your hair stands up. The rubbing causes some electrons in the atoms that make up your hair to leave their orbits in their originating atom. The balloon, made of rubber, is an insulator, and electrons that transfer to it remain bunched up on the balloon's surface. There is a transfer of electric charge from the hair to the balloon, creating a net positive charge in your hair and a net negative charge on the balloon. Since opposite charges attract each other, your hair stands up on end when the balloon is near. This phenomenon is an example of the EM force F_e, and it obeys a similar inverse square law to that of gravity:

$$F_e = k \frac{q_1 q_2}{r^2}$$

The values of the electric charge[6] of two objects, q_1 and q_2, are analogous to masses in the gravitational force. The EM force is responsible for chemical reactions and molecular bonding; in everyday life, this force completely overwhelms gravity. As the apple falls to the ground, the strong EM force acts to hold the apple's molecules together; the same forces acts on the molecules of silicon, holding dirt together. A dramatic event happens when the apple reaches the ground: the apple comes to an abrupt halt and is indented. If gravity were stronger than the EM force, the apple would keep going through the Earth toward its center.

The EM force is orders of magnitude stronger than gravity.[7] So why don't we see planets, stars, and galaxies clinging to each other as a balloon clings to a sweater? The reason is that most matter is electrically neutral: overall, the amount of positive charge equals the amount of negative charge in the universe. When a buildup of one charge occurs, the carriers of electric charge, generally electrons, quickly jump from one object to another to neutralize the charge. The spectacular event of a lightning strike reminds us of the enormous force associated with electricity.

Why are the strengths of gravity and electricity so vastly different? We don't know. But what we do know is that if their values and their differences were not as they are, we would not be here to talk about it. If gravity were just 5 % stronger, for example, our world would be very different. The Sun would burn its fuel considerably faster, causing it to be far hotter and therefore have a shorter lifetime. The Earth would orbit the Sun in a significantly tighter orbit (10 % closer to the Sun), and the Earth's core would compress. The internal heating would cause increased volcanism on the planet. Even a small change to gravity would have major consequences on our world, to the point that life could not have evolved here. Both electromagnetic force and gravity are "obligated" to have the strengths they do, no weaker and no stronger. According to Martin Rees, a prominent cosmologist, if this ratio were just a bit smaller, only a small and short-lived universe could exist.[8] It almost appears that these forces are "fine-tuned" to be what they need to be so that we can be here. However, the story does not stop there.

The other two forces of nature, the strong and weak nuclear forces, are far less known to most of us. Even their names are not representative of what they are. The weak nuclear force is actually very powerful compared with gravity, but it acts only within very short distances. This force is also orders of magnitude weaker than both the EM and strong nuclear forces. The weak nuclear force is responsible for radioactive decay. A neutron in free space, for example, will decay to a proton in just under 15 minutes. The same force is also responsible for *nuclear fusion*.[9] Energy, in the form of light, is liberated in this process and is ultimately responsible for heating the Earth and allowing life to exist. Carbon is also created in the fusion reaction of stars. The strength of the weak force ultimately determines how much carbon will be available. If the weak force were just a bit different than it is, insufficient amounts of carbon would be present for life to occur.

The behavior of the two nuclear forces also has puzzled physicists. For example, the energy associated with the strong force binds nuclei together. This force gives rise to most of the mass of the nucleons. Since nucleons represent most of the mass of the atom, the mass of almost all matter around us is ultimately dependent on the value of the strong nuclear force.[10]

The strong force also overwhelms all other forces of nature. To understand its strength, let us start with a comparison with the EM force. Imagine putting two protons next to each other. Particles with similar charge repel each other. So two protons, carrying a positive charge of 1.9×10^{-19} coulombs, placed at about one proton diameter apart, or 10^{-15} meters, would repel each other with tremendous force. How repulsive is this? Have you ever tried to force two bar magnets together, compelling similar poles to touch? Have you felt the magnets fight back? Now imagine a thimbleful of protons squeezed together in this fashion. To do this would require a force with the equivalent strength that would keep our moon hovering just a meter above the Earth's surface and not come crashing down.[11] But the strong force is 137 times stronger than the EM force. The EM force, like the gravitational force, gently decreases in strength with distance according to the inverse square law, or $1/d^2$. The strong force, unlike any of the other forces, increases in strength with distance. So strange is this situation that particles appear out of nowhere (out of the vacuum).[12] The weak force, in comparison, operates only in an exceptionally small range, virtually disappearing after 10^{-18} m or 0.1 % of the diameter of a proton.

Despite the tremendous differences among the four forces, they need to be as they are for us to exist. Interestingly, there are no physical laws that say these forces must be as they are. Several authors have made remarks on this seemingly strange situation of fine-tuning. One of the earlier remarks came in a well-known book on the subject in 1986 by Barrow and Tipler:

> A 50 % decrease in the strength of the [weak] nuclear force...would adversely affect the stability of all the elements essential to living organisms and biological systems. A bit more of a decrease, and there wouldn't be any stable elements except hydrogen.[13]

The Canadian philosopher, John Leslie, in his work *Universes,* and in his summary lecture, "The Prerequisites of Life,"[14] records the writings of several prominent scientists and their views of the strong force and its connection to life:[15]

> The strong force must be neither over-strong nor over-weak for stars to operate life-encouragingly. "As small an increase as 2 % in its strength would block the formation of protons out of quarks," preventing the existence even of hydrogen atoms,[16] let alone others. If this argument fails, then the same small increase could still spell disaster by binding protons into diprotons: all hydrogen would now become helium early in the Bang[17] and stars would burn by the strong interaction[18] which...proceeds 10^{18} times faster than the weak interaction which controls our sun. A yet tinier increase, perhaps of 1 %, would so change nuclear resonance levels that almost all carbon would be burned to oxygen.[19] A somewhat greater increase, of about 10 %, would again ruin stellar carbon synthesis, this time changing resonance levels so that there would be little burning beyond carbon's predecessor, helium.[20] One a trifle greater than this would lead to "nuclei of almost unlimited size,"[21] even small bodies becoming "mini neutron stars."[22] All which is true despite the short range of the strong force. Were it long-range then the universe would be "wound down into a single blob."[23]

59 | A leap in science

In the past decades, scientists have scrutinized how fine-tuning plays a role in terms of the weak nuclear force, especially in the role of *nucleosynthesis,* the process of creating atoms more complex than hydrogen. All the atoms in our bodies, including the oxygen that we breathe and the carbon in our bodies, all the iron in our buildings, and all the silicon in the Earth would disappear if the weak force were slightly different.

Fundamental Forces			
Force	Symbols*	Strength	Range (meters)
Strong		1	10^{-15}
Electro-magnetic		1/137	Infinite
Weak	neutrino	10^{-6}	10^{-15}
Gravity	m, M	6×10^{-39}	Infinite

* *Symbols:* • *neutron;* + *proton;* - *electron; m, M mass.*

Table of the relative strengths of the four fundamental forces in nature (Note: The "neutrino" will be revisited in Chapter 5.)

When two protons fuse, a fraction of energy is released in the form of light. That fraction is referred to as ε; it has a value of 0.007.[24] Though ε is small, the amount of energy released is still sufficient to make a star shine. For example, when four nucleons fuse to form helium, .007 or 0.7 % of their mass is converted to energy,[25] but that is still enough to effectively warm our planet, 150 million kilometers away, to a comfortable 15C (59F) average.

What if ε was different? What if it were 0.07, or .002 or .005 or 15 or 200,000? There seems to be no governing controller in nature that regulates what ε should be. But we know that the existence of life is highly dependent on and sensitive to its value. If ε were, for example, 0.006, hydrogen would not be able to fuse to produce helium efficiently; larger elements could not be created, and therefore complex chemistry and life would be impossible. According to Rees, if ε were above 0.008, all the hydrogen would have been fused shortly after the Big Bang. Insufficient hydrogen would have been left over for stars to produce energy and life-essential carbon. Some physicists disagree on the exact value of ε, but all agree that it cannot change appreciably for life in the universe to exist.

Fine-tuning and the miracle of carbon production

The most basic building block for life is carbon, and this is due to its two distinctive properties. Carbon is a light element, allowing it to form substances that are either airborne or soluble in water and, therefore, efficiently transported. Carbon also has tetrahedral bonding, allowing it to connect with elements and compounds in four directions. In other words, carbon is the ultimate Lego piece that can attach itself to other atoms more easily than any other, and because it's not too massive, it can be brought to its needed location with ease and a little water. This characteristic makes carbon central in the development of large and complex molecules. Scientists have considered other elements as building blocks for life as well. The next lightest element in the carbon group is silicon. Some have speculated that since silicon's outer electron shell, consisting of its *valence electrons*, is similar to that of carbon, silicon could replace carbon as an alternative life-element. As silicon is a common element in the cosmos, this would increase the chances of detecting life elsewhere. However, the problem with silicon is that it's a heavy element. The additional electron shell in silicon means that its valence electrons form weak bonds with other elements—it's a bad Lego piece. The result is that silicon forms sand, not living tissue. Fortunately, carbon

61 | A leap in science

is a common element in the universe; it may be hard to imagine, given the above discussion, that its abundance is a result of fully random processes. But if you are not convinced yet, you may be when you see how carbon is produced.

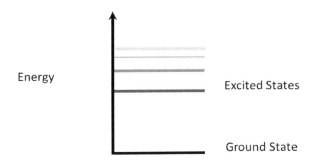

Energy levels in a nucleus including ground and excited states

Stellar nucleosynthesis is the process of creating new elements via the nuclear fusion that occurs in stellar cores. This process has occurred in stars for billions of years and is the reason the universe is filled with elements more massive than the primordial elements hydrogen and helium. Carbon is one of those massive elements, and it's fascinating to see how it comes into existence.

Carbon-12 is a stable isotope[26] and stars abundantly produce it; the collision of beryllium-8 and helium-4 produces this important isotope. The strong, weak, and EM forces must work together for this to happen. Many types of particle collisions occur every second in the stellar fusion process. But one set of collisions happens when two helium-4 nuclei, composed of two protons and two neutrons, collide and create the isotope beryllium-8, with four protons and four neutrons. For every proton added to a nucleus, a neutron is required to counterbalance the powerful repulsive EM force. Imagine a whole bunch of social introverts at a party, repelling each other. To get people to stay and the party to continue you would need a few extroverts mixed in. So as the size of the nucleus grows, nature desires additional neutrons to offset the growing instability of many protons crammed together in the nucleus. Beryllium, for example, typically has four protons and five neutrons, and is called, naturally, beryllium-9. An isotope of beryllium, beryllium-8, is created by the collision of two helium-4, and if beryllium-8 collides with another helium-4 nucleus, it will become carbon-12.

There are subtleties in nature at the nuclear scale that benefit carbon formation; this will require a short venture into quantum mechanics and water droplets. Particles can absorb energy from radiation[27] and be "excited." In other words, particles are in elevated energy states, above their lowest energy state, or ground state; see the schematic above[28]. When a helium nucleus (also known as an alpha particle) hits a beryllium-8 nucleus, the combined new composite is carbon-12. The collision is less like two billiard balls colliding and more like two water droplets hitting each other. The particles coalesce, rotate around each other, then separate, and possibly break up into two or more particles. Though subatomic particles follow different force rules than water droplets, their behavior in the collision has similar characteristics. The strong and EM forces each play a role here. Both the beryllium and alpha particles are composed of protons and neutrons, and as we have already seen, each of these is formed from subnuclear particles called quarks. When nucleons collide, there is no clean collision; quarks, interacting by the strong force, interact with each other. Nucleons exchange quarks with other nucleons all the time. When beryllium-8, with eight nucleons, hits an alpha with its four nucleons, the result is a complicated multiparticle collision.[29] Like water droplets colliding, there is squeezing, rotating, splitting, and reforming that takes place (see figure on the following page[30]). If that analogy doesn't help, imagine that two people, each with a grocery bag and walking at a brisk clip, approach each other blindly at a building corner. Now imagine that each sack is full of delicate and thin jars of honey. The collision would result in ripped bags, mixed honey and two frustrated (excited!) and messy individuals that need some time to clean up.[31] For the colliding alphas, the collision is also a sticky mess, and before the new beryllium-8 settles to a stable state, the temporary excited state exists—the amount of time beryllium remains excited is critical.

Carbon production requires that the frequency of collisions is fine-tuned; not too frequent nor too infrequent. Beryllium-8 is unstable and decays or breaks apart. If beryllium hits another particle, other than an alpha particle, no carbon is formed; if it decays too quickly, again, no carbon is formed. But fortunately, beryllium lives long enough in the formation process described above that there is a small probability that it may enter into another collision with an alpha and therefore make carbon-12.[32] When the beryllium-8 and the alpha collide, they bring with them kinetic energy that subsequently must go somewhere. Quantum mechanics says that at the microscale, only specific energy slots are available. When two particles collide, the resulting particle must have a particular energy slot available where that incoming kinetic energy can go (as seen in the schematic). If no energy slot is available, the outgoing particles will not stick together.

Schematic of water droplets colliding (Source: ResearchGate.net and adapted by C.C. Zachary)

Before 1952, no such energy state was known that would permit the fusion of beryllium-8 and helium-4. These two nuclei were not expected to stick together to make carbon-12. They should have departed from the collision as two or more separate entities. The astrophysicist, Fred Hoyle, reasoned that the abundance of carbon-12 in the universe meant there must be an undiscovered energy level for carbon-12. Analogous to "sticky" water in the water droplet example, this yet-to-be-discovered energy level would allow for carbon-12 to "stick together" and explain the abundance of this element. Hoyle convinced a colleague to run an experiment to find this excited state—and they quickly discovered the new energy level.[33] This excited energy state, also known as the *Hoyle energy state*, helped to properly account for the amount of carbon found in the cosmos. This was a significant discovery and one of the first to use the *anthropic principle*[34] to make scientific progress.

The entire scenario just described is one of the most stunning examples of fine-tuning.[35] Since no natural process governs all these independent processes, we can rightly ask: Is this entire scenario just a large string of coincidences?

A fine-tuned cosmos

Fine-tuning not only shows up at the nuclear scale, but it also appears at the cosmic scale. The average density of the universe, symbolized by Ω (the Greek omega), represents the ratio of the actual density of the universe to the critical density or the density that would cause the universe to collapse. If Ω were too large, the universe would have collapsed long ago; if Ω were too small, the universe would have expanded too fast for matter to coalesce and form galaxies and stars. In either case, Ω must be fine-tuned and indeed, it may well be.[36] The universe is expanding at just the right rate to allow for galaxies to form and to have allowed enough time for appropriate stars and planets to evolve so that life could emerge.[37]

Scientists see fine-tuning in the way our universe is expanding. Imagine detonating an underground bomb; the explosion throws dirt and rocks into the air. After a few moments, the material is pulled back to earth by the force of gravity. After matter and space were created at the moment of the Big Bang, everything emanated out from a point. And just as rocks are thrown up in the air and eventually fall back to earth, it's expected that the universe would expand and ultimately slow in its expansion. One scenario is that the universe would even "fall" back to a single point (the Big Crunch). But in the 1990s a surprising discovery was made.[38] Space appears to be expanding faster and faster, as if the dirt in our explosion was attracted skyward by some mysterious force! As odd as this may seem, there appears to be a mechanism in the universe that is pulling things apart.[39] Though this force is not completely understood, Albert Einstein inadvertently accounted for this force in his adjustment to general relativity. He purposely inserted this term, known as the cosmological constant λ (the Greek lambda) to create a steady-state universe. This "fudge factor," as it was colloquially called, was designed to make Einstein's famous field equations fit the mainstream philosophy of the early twentieth century—that the universe was infinitely old. Einstein was unable to believe the simplicity of what his equations were telling him: that the universe was evolving and had a beginning. The term λ artificially forced his equations to match that of an eternal universe.

A leap in science

Though the original formulation of general relativity included only the attractive form of gravity, this new term represented a repulsive force. Einstein could use the attractive and repulsive forms of gravity to tune or balance one another, resulting in a permanent and unchanging cosmos.[40]

The wrong notion of an infinitely old universe temporarily misled the great scientist. With the construction of the Hooker Telescope on Mt. Wilson, Einstein realized his misconception of the cosmological constant.[41] In 1929, only a little over a decade since the cosmological constant was introduced, astronomer Edwin Hubble made one of the most important discoveries of the twentieth century: the universe is expanding.[42] The cosmological constant was no longer needed. But decades after λ was removed from the field equations, it was revived in the 1990s. Observations have shown that the universe is apparently expanding and even accelerating in its expansion, and the additional term accounts for this otherwise unexplained phenomenon. Though the actual value of the constant is still in debate, scientists understand that it must be small. If it were much greater than about 10^{-122} the repulsive force would prevent galaxies and stars from forming. However, if λ were much smaller, the observed acceleration in the expansion of the universe could not be explained.

The problem with fine-tuning λ is very much akin to the fine-tuning of Ω, the critical density parameter, but in this case, fine-tuning is extreme. Mathematicians sometimes use grains of sand on a beach to describe huge numbers, or inversely, very small probabilities. One grain of sand divided by all the grains of sand on earth doesn't even come close to describing the incredible precision λ must have for it to fit observation. Now imagine if every planet of every solar system in the universe were filled with sand and one grain was set apart. That one grain divided by all the other grains would produce a small ratio that still wouldn't come close to the sensitivity of λ. We only approach how sensitive this number is if we imagine every cubic meter of space filled solid with protons (each occupying about $10^{-45} m^3$) and every cubic meter of the observable universe (about $10^{80} m^3$) filled similarly. Now imagine one proton is picked out of 10^{125} protons. The difference that single proton makes out of that astronomical number would result in a universe very different, where life could not exist.[43] The universe we live in is astoundingly fine-tuned for our benefit.[44]

Beyond fine-tuning, there are other unexplainable realities concerning our origins. How did it all begin? What would have been required for everything to just "come together" and spontaneously explode? Remember, as far as we know, nothing existed before the Big Bang, so something had to be created out of nothing, absolutely nothing.[45]

Some theorists have clever ways to deal with the "nothing" before the Big Bang. Some models explain that "something" or at least a principle (the uncertainty principle[46]) existed before the Big Bang. Still, the scientific community has no consensus on this. Furthermore, it's impossible to test a principle or theory before time existed.

One of the significant problems for randomness to be responsible for the Big Bang has to do with *entropy*. Entropy is a thermodynamic term related to energy; it quantifies how well energy is organized. The more entropy, the more energy is disorganized; the less entropy, the more energy is organized. Entropy can also be visualized in the following way: Visualize putting in the effort (and therefore the energy) to clean your child's room in the morning. You know where things in the room go, and you make a concerted effort to arrange objects as they should be. Unless you have children who know the rules and perhaps have some incentive to apply entropy-reversing procedures, the fundamental laws of physics will take over at that point.

Now imagine two pairs of socks arranged on the bed. Unless we apply a bit of care (and energy) to organizing the socks, one sock will pair with a wrong one and then we will have two mismatched pairs. The chance that a correct pairing will be done haphazardly or randomly is 33 %.[47] If three pairs are laid on the bed, the chances that each sock will be paired correctly with the right one drops from 33 % to one chance in five or 20 %. The rest of the room will suffer from similar odds; things naturally get messy. Depending on the room's inhabitant, in a few hours or days, that room will be disorganized: the bed will be unmade; toys and clothes will be on the floor. The room will become, as we say in physics, chaos.

Our universe is undergoing a similar process; it's becoming a vast "messy bedroom." Over time the thermodynamic energy of the universe is winding down, losing heat (eventually leading to *heat death*), and entropy is increasing on a massive scale.

67 | A leap in science

Ironically, it appears to us that the universe near us is organized and therefore even decreasing in entropy. For life to exist, elements must be arranged in very complex molecules, and even more complex cells and higher life forms. But this irony is explained if you look deeper into the problem. Plants, for example, take in energy through photosynthesis; a relatively low number of high-energy photons from the Sun are absorbed and export energy with relatively many photons of lower energy. Entropy reduces for the plant at the expense of increasing entropy of the photons that supply energy to it. Overall, entropy increases. This same process applies to any form of life, whether it be a plant using energy from photons, or animals using the chemical energy from plant intake. As a matter of fact, any process in the universe that decreases entropy locally is also responsible for increasing it globally.

Since entropy is overall increasing, the universe must have been in a lower state of entropy earlier in its existence, and at its lowest state at the very beginning. This brings us to a fundamental question: What are the chances that the universe could have started in its lowest entropy state? In other words, before matter existed (before time began), what are the chances that random fluctuations of energy from nothingness organized themselves to create everything we know? This is like asking what the chances are that our socks, when thrown on the bed as they came out of the dryer, paired themselves in the right way, without any careful attention (or energy) applied. We have shown that we can calculate the chances for a couple of pairs of socks, but how about the universe? Is the chance as small as one in a million? One in a billion? One in a trillion? Not even close.

Rodger Penrose, the Rouse Ball Professor of Mathematics at the University of Oxford and a good friend of Stephen Hawking, studied this very question. Begin with the total number of particles in the known universe. For this discussion, we can assume this is about 10^{80} particles. (This number may be significantly underestimated, and if this is the case, the result is even more spectacular.)

According to Penrose, the chance for the universe to appear is related to, in a certain sense, the number of possibilities for it not to appear.[48] We have here an interesting counting exercise, but instead of socks, we are talking about *phase space*, the set of possibilities of position (space) of particles and their movement.[49] The number of possibilities for each particle can be found and can be used to give us the probability of them being in their lowest entropy state.[50] The chance is astonishingly small, one in 10, raised to the power of 10, and again raised to the power of 123 or $10^{(10^{123})}$ or:

$$p = 1 / 10^{10...0}$$

—where we have left out 121 zeros.

The denominator of this probability is so large that it cannot be written down. The fantastically large number, the googol (10^{100}) and the larger number, the googolplex (10^{googol}) do not even come close to the denominator in probability p above. If each number in $10^{10...0}$ was written down with the average font size for a book, this number could not fit inside the known universe. The probability that our universe is here by chance is not just small, it's unfathomably small.

Given the arrays of probabilities in this chapter, many scientists argue that we simply do not know enough to answer all the scientific questions that arise (I agree) but that science will eventually give us a complete explanation (that, I doubt). The great cosmologist, John Wheeler, collected these thoughts on this subject and eloquently put it this way:

> To my mind, there must be at the bottom of it all, not an utterly simple equation, but an utterly simple *idea*. And to me that idea, when we finally discover it, will be so compelling, and so inevitable, so beautiful, we will all say to each other, "How could it have ever been otherwise?"[51]

Is it reasonable to take a step of faith toward the "idea" that John Wheeler mentions? The question is: What idea? Maybe, just maybe, that idea transcends science.

Chapter summary and questions

Our universe is a well-balanced place! The four forces (gravity, electromagnetism, the weak nuclear force, and the strong nuclear force) found in nature are "fine-tuned." If any of them were a bit different in strength or had a different behavior over distance, it would not be possible for life to exist in the universe. Beyond these forces, our very existence is a "miracle." Energy would have to be so organized (low entropy) at the beginning, that no credible idea currently exists (or can be tested) to account for why we are here. Hence, it takes abundant faith to believe that the creation of the universe came about by chance.

Questions to ponder:
- Do you have examples in your everyday life of things that are fine-tuned? How are they fine-tuned and what is responsible for them being so? (Hint: think of your kitchen or kid's bedroom. These rooms are useful only if things are correctly arranged in them.)
- Have you ever had several bad things happen independently in one day (e.g., flat tire, being fired from your job, another flat tire coming home, finding your house burned down when you finally returned home)? Fortunately, the chances for any of these events happening on any one day is small. So how much less probable would it be for all of them to occur on the same day? How would you rationalize several bad (or good) things happening in the same day? Do you have an example?
- Do you have an example of entropy being reduced in your everyday experience? If yes, did you experience any energy needed to accomplish this? Hint: Think about cleaning your house.

NOTES
[1] Quantum mechanics tells us that small subatomic particles don't actually move in orbits but are energy waves that have no distinct position. Their location can only be identified with a certain probability.
[2] Physicist and Nobel laureate Murray Gell-Man is believed to have named this elementary particle. He started by using names like "squeak" and "squork" for peculiar objects, and "quork" (rhyming with pork). Some months later, he came across a line from James Joyce's *Finnegans Wake* where quark was used (in a completely different context); Gell-Man liked that term and it stuck.
[3] Martin Rees, *Just Six Numbers: The Deep Forces That Shape the Universe*, 1st American ed. (New York: Basic Books, 2001), 4.
[4] This is the classic picture of gravity, but it is only an approximation: general relativity says that space and time are distorted by mass, creating an effective force field with a geometric interpretation.
[5] The universal gravitational constant is in units of mass cubed, inverse kilograms, and inverse squared seconds, or $G = 6.674 \times 10^{-11}$ $m^3 \cdot kg^{-1} \cdot s^{-2}$.

[6] The electromagnetic force constant k, or Coulomb's constant, has the value of 8.99×10^9 Nm²C⁻² in metric units of force or newtons, meters squared, and coulombs (unit of charge).
[7] Two electrons, each with a mass of 9.1×10^{-31} kg and charge 1.9×10^{-19} C, are placed a meter from each other, $r = 1$ meter. The ratio of the two forces, electric and gravitational is then: $F_e / F_g = k\ (q_1\ q_2)/r^2 / G\ (m_1\ m_2)/r^2$. Simplifying and plugging in values for k, G, q and m, $F_e / F_g = 4.1 \times 10^{42}$. The electromagnetic force overwhelms gravity!
[8] Rees, *Just Six Numbers*.
[9] Fusion occurs when light nuclei combine with other nuclei; for example, the hydrogen nucleus (a single proton), combines with another hydrogen nucleus and creates a helium nucleus. Energy is liberated in the process.
[10] Energy and mass are interchangeable, according to Einstein's famous relationship, $E = mc^2$. Individual quarks, the constituents of nucleons, provide only about 1 % of the mass of a proton. The complement of the mass comes from the energy responsible for holding the quarks bound, the *binding energy*. Also, a proton is 1835 times more massive than an electron.
[11] Of course, the direction of the force in this example is the opposite: protons repel each other, while the Earth-Moon force is attractive.
[12] This increase in strength with distance means this force behaves in a nonintuitive way. When two particles, attracted by the strong force, are pulled away from each other, new particles are created out of the vacuum (essentially from nothing) to dissipate this force.
[13] J.D. Barrow and F.J. Tipler, *The Anthropic Cosmological Principle* (Oxford: Oxford University Press, 1986), 327.
[14] J. Leslie, "The Prerequisites of Life" in *Newton and the New Direction in Science*, eds. G.V. Coyne, M. Heller and J. Aycinski (Vatican City: Specola Vaticana, 1988), Section II.
[15] J. Leslie, *Universes* (New York: Routledge, 1989), 35–36.
[16] J.D. Barrow and J. Silk, "The Structure of the Early Universe," *Scientific American* 242 No. 4 (1980), 127–8.
[17] P.C.W. Davies, "The Anthropic Principle," in *Particle and Nuclear Physics* 10, 1–38, 1983, found in F.J. Tipler, *The Anthropic Cosmological Principle* (Oxford: Oxford University Press, 1986).
[18] F.J. Dyson, "Energy in the Universe," *Scientific American* 225 (1971), 52–4 and Idlis, Izvest. *Astrofiz. Instit.* Kazakh. SSR 7 (1958), 39–54, esp. 47, 56.
[19] F. Hoyle, "On Nuclear Reactions Occurring in Very Hot Stars," *Astrophysics. J. Suppl.* 1 (1954): 121. E.E. Salpeter, "Nuclear Reactions in Stars," *Physical Review* 107 (1957), 516.
[20] I.L. Rozental, *Structure of the Universe and Fundamental Constants* (Moscow: 1981), 8.
[21] B.J. Carr and M.J. Rees, "The anthropic principle and the structure of the physical world," *Nature* 278 (1979), 611.
[22] B. Carter, *Atomic Masses and Fundamental Constants: 5* (New York: 1976), eds. J.H. Sanders and A.H. Wapstra, 652. 42.

71 | A leap in science

[23] P.W. Atkins, *The Creation* (New York: W.H. Freeman, 1981), 13.
[24] This same energy production is what makes hydrogen bombs so explosive.
[25] Rees, *Just Six Numbers*.
[26] An isotope is a variation of an element with the same number of protons and electrons, but varying neutrons; the sum of protons and neutrons gives the atomic mass number N, written after the name of the element in this text.
[27] Radiation is in the form of high-energy light, gamma rays, and X-rays.
[28] Energy states are a bit like the rungs of a ladder. Energy can be absorbed or transmitted from an atom (or particle) in discrete units or steps. You can't have energy absorbed or transmitted if it doesn't come from one of its defined states, just like you can't step between the rungs of a letter and expect to climb.
[29] In addition, each nucleon contains three smaller subatomic particles, quarks. Altogether, there are potentially $3 \times 8 + 3 \times 4 = 36$ particles interacting (it's a mess!)
[30] ResearchGate.net: "Separation collision of two 300 μm water droplets."
[31] This actually happened when I was a graduate student on one of my late-night shopping excursions, honey excluded.
[32] Two good references on the subject are: M. Livio, D. Hollowell, A. Weiss and J. W. Truran, "The anthropic significance of the existence of an excited state of 12C," *Nature* 340:6231 (27 July 1989): 281–284, and Jean-Philippe Uzan, "The fundamental constants and their variation: Observational and theoretical status," *Reviews of Modern Physics* 75:2 (April 2003): 403–455.
[33] This excited state, known as the Hoyle state, is one of the most astonishing discoveries in stellar nucleosynthesis. For more detail, see Helge Kragh, "When is a prediction anthropic? Fred Hoyle and the 7.65 MeV carbon resonance," 2010, http://philsci-archive.pitt.edu/5332/
[34] The anthropic principle: the cosmological principle that theories (and conditions) of the universe are constrained by the necessity to allow for life and, ultimately, higher intelligent life.
[35] Not all physicists are convinced of the fine-tuning argument found in the Hoyle state or any other process described in this chapter. One possible alternative mechanism is the multiverse theory, but more will be said on that in Chapter 9.
[36] Some argue that the inflationary theory of the universe (see Chapter 9) provides a solution as to why $\Omega \approx 1$ (flatness problem). But the theory is not free from issues and does not, on its own, solve the flatness problem.
[37] Rees, *Just Six Numbers*.
[38] These observations of an accelerating expanding universe were made by Adam Reiss, Saul Perlmutter, and Brian P. Schmidt, cowinners of the 2011 Nobel Prize in Physics.
[39] Currently, dark energy (suggested to be the vacuum energy of space—particles pulled out of the vacuum due to the Heisenberg uncertainty principle) is thought to be responsible for the universe being in an accelerated expansion phase. Dark energy is thought to permeate all of space. Dennis Overbye, "Cosmos Controversy: The Universe Is Expanding, but How Fast?" *The New York Times*, 20 February 2017, retrieved 21 February 2017, and P. J. E. Peebles and Bharat

Ratra, "The cosmological constant and dark energy," *Reviews of Modern Physics* 75:2 (2003): 559–606.

[40] D. Lincoln, "Einstein's True Biggest Blunder (Op-Ed)," Fermi National Accelerator Laboratory, November 6, 2015, received from http://space.com, December 8, 2017.

[41] Public Broadcasting Station (PBS), "Cosmological Constant," PBS.org, retrieved May 29, 2011. It was reported by a colleague, George Gamow, but never verified, that Einstein said that changing his equations was "the biggest blunder of [his] life." Ironically, it turns out that the addition of the term is actually needed, but for a different reason.

[42] The expansion was observed looking at distant galaxies and recording their recession velocities. Hubble was able to connect the distance of these objects and their recession velocities and create a simple law that showed that the farther each galaxy was from us, the faster it was receding from us, much like particles in an explosion that emanated from a single point. Since the universe has no special point, it was understood that space itself was expanding. The entire universe must have begun in some huge explosion, later called the Big Bang, a term coined by the astrophysicist previously mentioned, Fred Hoyle.

[43] Or by adding negative exponents, $10^{-45} \times 10^{-80} = 10^{-125}$.

[44] According to Rees, there are other fined-tuned quantities in the cosmos that must be in place for life to exist; some conditions may seem obvious, others not. Consider the fact that we live in a universe that has three dimensions of space and one dimension of time. Had there been some other arrangement, like four spatial dimensions, the additional freedom for particles to move around would not allow for the needed subatomic process to carry through and, again, the possibility of life would not exist.

[45] Things get very tricky here: both general relativity and quantum mechanics are needed to understand the physical nature of matter when the universe was very young and small, much smaller than an atomic nucleus. The fundamental relationship of uncertainty by Heisenberg may imply that a perfect vacuum may not exist, meaning that there was a sort of cosmological constant before the Big Bang. If this is true, then it is possible that a field of energy could create a particle and antiparticle creation. That "particle" and "antiparticle" would be our entire universe and, supposedly, an anti-universe. There is a good deal of speculation here, and a theory of quantum gravity would be required to understand this. This idea also presumes that the uncertainty principle exists before the beginning—something that can never be tested.

[46] The uncertainty principle, or the Heisenberg uncertainty principle, was articulated (in 1927) by the German physicist Werner Heisenberg: the position and the velocity of an object cannot both be measured exactly, at the same time, even in theory. This principle essentially says that "exactness" in nature has no meaning.

[47] There are three socks that can be picked in a random fashion, one correct one and two incorrect. Pairing correctly has 1 chance in 3 (33 %) and incorrectly 2 chances in 3 (66 %).

[48] Rodger Penrose, *The Emperor's New Mind: Concerning Computers, Minds, and the Laws of Physics* (New York: Viking Penguin, 1990), 339–345.
[49] And movement is expressed in these terms: momentum is a product of mass and velocity.
[50] According to the calculations of Jacob Bekenstein and Stephen Hawking, maximum entropy for particles occurs when matter is compressed to its extreme, in a black hole. Jacob D. Bekenstein, "Black holes and entropy," *Physical Review D*. 7:8 (April 1973): 2333–2346, and Stephen W. Hawking, "Black hole explosions?" *Nature* 248:5443 (1974): 30–31.
[51] John Wheeler, former chair of the physics department at the University of Austin, quoted from the PBS science documentary, *The Creation of the Universe*, 2004.

Chapter 5 – Our Earth: A Unique Paradise or One of Many?

Look again at that dot. That's here. That's home. That's us. On it everyone you love, everyone you know, everyone you ever heard of, every human being who ever was, lived out their lives. The aggregate of our joy and suffering, thousands of confident religions, ideologies, and economic doctrines, every hunter and forager, every hero and coward, every creator and destroyer of civilization, every king and peasant, every young couple in love, every mother and father, hopeful child, inventor and explorer, every teacher of morals, every corrupt politician, every "superstar," every "supreme leader," every saint and sinner in the history of our species lived there—on a mote of dust suspended in a sunbeam. – Carl Sagan, Pale Blue Dot: A Vision of Human Future in Space *(1994)*

Carl Sagan memorialized the Earth in his epithet message of the Voyager spacecraft as it took its final images of our planet. On that one speck of faint blue light is us: all that humans have ever done, thought, or contemplated in all history, for all time. Except for a few dozen people who have ventured into space, our entire existence has been here on this blue, watery planet we call Earth. Our world is a "heaven" compared to every other place in the known universe, at least from what we know at this time. But our paradise did not just happen overnight. Science tells us that our home, the earth, took billions of years to be prepared. In this very unexceptional place in the vast universe, it was here where life sprung. That life eventually became self-aware and contemplated the creation and its Creator. It was here where humans evolved and contemplated thoughts never before considered. From a spiritual perspective, King Solomon summarized it this way: "He has made everything beautiful in its time. He has also set eternity in the human heart" (Ecclesiastes 3:11a NIV).

The Earth as seen from the Moon; the Apollo 11 Mission (Source: NASA)

Design and the anthropic principle

The great French philosopher, mathematician, and scientist, René Descartes, made the famous statement,

> The physical laws he had uncovered revealed the mechanical perfection of the workings of the universe to be akin to a watchmaker, wherein the watchmaker is God.[1]

The idea that the universe has been assembled through a thoughtful process and is orchestrated by an all-powerful being does invoke theology, but Descartes was not alone. Half a century later, Isaac Newton had this same perception of our world. And then in 1802, the English theologian William Paley made his famous watchmaker analogy. Paley pointed out that the universe has a design and so, logically, it must have a designer. The analogy is simple: Imagine finding an intricately designed watch on a path in a forest. We would never consider that the watch was produced by chance by some random mechanism of the forest. Neither should we consider our existence on planet Earth to have resulted from random happenings of the cosmos.[2]

The previous chapter explored the undeniable fine-tuning of the universe: if physical constants and processes were just a bit off, we would not be here to talk about it. The anthropic principle takes this one step further; it states that the universe must be as it is, with all its fine-tuned characteristics, fundamental constants, forces, structure, age, etc., because conscience and sapient life observes it.[3] There is no doubt the principle takes us in a step toward the direction of God. On the other hand, you cannot falsify the anthropic principle; you can neither disprove it nor prove it. We cannot test a batch of other universes to see if they have different constants, forces, structures, and age and then check to see if life appears. The anthropic principle is, therefore, not a scientific principle. Nonetheless, it's a powerful tool we can use in our discussion about our existence in the cosmos.

There are many versions of the anthropic principle, but astrophysicists Brandon Carter and Robert Dicke have suggested one workable version: only in a universe capable of eventually supporting life will there be living beings capable of observing and reflecting on the matter.[4] When applying this principle to what we observe, we can then ask important questions. In the words of the physicist-atheist turned believer, John Clayton,

> When we look at the universe as a whole, and we look at what it takes to sustain life, is chance a viable mechanism to explain the universe that we see and the Earth that we live on?[5]

Clayton's question is a fundamental one about life and how it arose. Life cannot appear on just any planet, around any star, or at any place in our galaxy; many restrictions dictate where life can begin, flourish, and ultimately give rise to a higher, self-aware form of life. The Earth is one such place, but what are the odds for another planet such as ours to exist? How unique is our Earth? We started with the argument of fine-tuning of constants and forces; now we make that same journey through the cosmos and eventually to our home. What we will find is truly amazing.

Our own neighborhood in the cosmos: The Milky Way

Our neighborhood, the Milky Way, is located in a quiet place on the outskirts of a cluster of galaxies.[6] Our own galaxy, known as the Milky Way, is a fabulous collection of more than a billion stars.[7] So vast is the Milky Way that the units of distance, the mile or the kilometer, are not useful here. The *light-year* (lyr), on the other hand, is a measure of the distance that light travels in one year, about 9.5 trillion kilometers or 5.9 trillion miles. The Milky Way is about 100,000 light-years across, and we are some 26,000 light-years from its center. Large spiral arms made of gas and dust extend from the circular bulge in the center of the Milky Way, giving our galaxy the classification of a spiral galaxy. More precisely, our Milky Way is a barred-type spiral: spiral arms extend from the center, attached by a bar-shaped formation composed of stars. Looking from "above" our galaxy would look like a similar galaxy (NGC 1300) shown in the figure on the following page.

It would take a spacecraft traveling at the speed of light (186,000 mi/sec) more than 26,000 years to voyage from our Earth to the center of our galaxy. The light that we see at the center of the galaxy[8] left its star when ice covered a significant portion of North America, Europe, and Asia in the last Ice Age. Our entire solar system is but a grain of dust in a vast expanse of stars that stretches beyond the imagination. As an analogy, if somehow one could compress the solar system and place it in a pillbox the size of a US quarter, the Milky Way would be the size of the continental United States.

As large and magnificent as our Milky Way is, it's an incredibly inhospitable place. For another solar system to have a planet that could support life, it must be in a particular location within the galaxy. There is a region called the *galactic habitable zone* (GHZ), a double annulus region, one ring above the galactic plane and one below. This is the region where astrobiologists and planetary astrophysicists believe that a planet orbiting a star could most likely harbor life. The GHZ is a place where the needed ingredients for life are found, including heavy elements and metals, but it's far enough from the dangerous areas of the galaxy where potential life would be destroyed (see figure below). The GHZ is outside the central bulge. This central region of the galaxy is an inhospitable place full of radiation, stellar collisions, and *supernova*.[9] We expect *black holes*[10] to be abundant in the galactic center, and any solar system found there would be exposed to high radiation coming from these objects. The proximity of stars in the bulge would cause significant fluctuations in gravitational forces from nearby stars, disrupting solar systems. The result is unstable planetary orbits and an environment inhospitable to life.

The New Galactic Catalogue (NGC) object number 1300, a barred galaxy that resembles our own Milky Way (Source: NASA, ESA, and The Hubble Heritage Team STScI/AURA)

Many of the galaxy's spiral arms are also inhospitable for a life-bearing planet. Our star, the Sun, was probably born in a stellar "incubator" along with significantly more massive stars. These star formation nurseries, called *nebulae*, offer some of the most spectacular sights in the heavens. The Eagle Nebula[11] is a brilliant example of stellar formation. Only a small part of this nebula (less than 1 % of the total) is shown in the figure, yet even the head of the Eagle is approximately 1000 times the size of our solar system. This space cloud, made of gas and dust, is 7000 light-years from us.

The Earth | 79

Gigantic stars, some thousands of times more massive than our sun, are forming there. But the proximity to stars in this nebula, some producing millions of times more light than our sun, makes conditions unsuitable for life on any planet orbiting any of its stars. Deadly radiation from nearby young stars is abundant there. Giant stars, whose surface temperatures are 10 times hotter than our Sun and which are more than a million times more luminous than the sun, produce radiation that would penetrate all but the densest planetary atmospheres, killing any life that might be there. And similar to the galactic bulge, tidal forces resulting from close encounters with other stars tend to destabilize planets in their orbits, causing some planets to leave their stellar host, and others to plunge toward their star. Other galactic star clusters, *globular clusters*,[12] are also inappropriate for life for similar reasons.

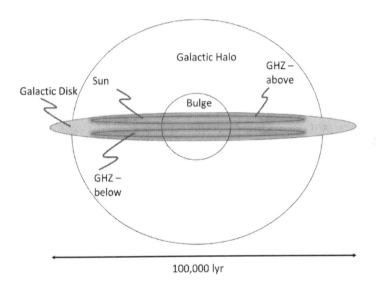

Sketch of the galaxy and the galactic habitable zone (above and below the galactic equator)

For life on a planet to begin, evolve, and achieve self-consciousness requires billions of years under relatively "calm" conditions. As we have seen, a significant portion of the Milky Way is not calm. As we go toward the outer half of the galaxy, we find places that are more placid and stable, but they too are considered uninhabitable, for different reasons. The outer parts of the Milky Way lack the massive star formation found in the galaxy's arms, and therefore heavy metals, needed for life, are deficient there.

There are other types of galaxies, quite different from our Milky Way. According to the Hubble Space Telescope survey, roughly 72 percent of the galaxies observed are spirals.[13] Though most of our galaxy is inhospitable to life, from what we know of other galaxies, they are at least as unfriendly.[14]

An elliptical galaxy in the foreground of a cluster of spiral galaxies (Source: NASA and ESA)

The lives of stars and their connection to us

Ironically, for life to exist, a certain amount of "controlled violence" is needed. Spiral galaxies offer what appears to be the right balance for providing a star system with the necessary violent start followed by an extended period of tranquility.

When the universe was very young, only hydrogen and helium, the two lightest elements, were present. But life could not exist with only these two light elements. Hydrogen cannot form complex molecules needed for life, and helium is inert and therefore does not form molecules.

Heavier elements such as oxygen, nitrogen, and primarily carbon are all required for life to exist. Within the cores of stars, in the inferno of nucleosynthesis, these elements were produced. Even heavier elements, such as iron and nickel, are also crucial for life, and these elements are created in the centers of more massive stars. This entire process all happens in a short time at the end of the massive stars' lives.

No star lives forever. Since the Big Bang 13.8 billion years ago, multitudes of stars were born and have died. The larger the star, the brighter it shines and the shorter its life. Our star, the Sun, is a yellow dwarf; it's midrange in mass and "calmly" converts hydrogen to heavier elements in the fusion process over its long life. Fusion produces pure energy of high-intensity radiation. The core of the Sun is blazing at 15 million degrees Celsius or 27 million degrees Fahrenheit! This is hot enough for hydrogen to fuse, and through several intermediate steps, this lightest element is converted into carbon and oxygen. For six billion years, this process has been going on, and it will continue for another six billion.

Our Sun and solar system were probably born in a much more violent place than the quiet patch of the galaxy we find ourselves in today. Such a site might have looked something like the Eagle Nebula in the figure on the following page. These places are ideal for stars and planets to begin their lives, since the clouds are filled with the material to create a star: the ghostly remnants of exploded stars. All naturally occurring elements are found here, including hydrogen, carbon, nitrogen, oxygen, and heavier metals such as nickel and iron, to name a few. Though the core of the Sun is unimaginably hot by Earth standards, it's not massive enough to produce heavy elements; heavy element production is left to more massive stars.

Stars greater than about 25 solar masses (25 times the mass of our sun[15]) are capable of producing all the elements; these stars also undergo an incredibly violent termination of their short lives. Stars eject material from their interior and thrust it into space where it can accumulate in clouds. One day, these same clouds will give birth to another solar system like our own.

Pillars of creation: a small section of the Eagle Nebula and where stars are born (Source: Arizona State University, NASA, ESA, and The Hubble Heritage Team STScI/AURA)

The lifetime of a 25-solar-mass star is significantly shorter than the Sun's. A massive star such as this will burn through its fuel so ferociously that it will consume it in less than 7 million years. In astronomical terms, this is a blink of an eye. Each time the star consumes a light element, a heavier element is then fused to make an even heavier one. At each step, the star extracts energy, and that energy keeps the star from collapsing from its own weight.[16] Once the core has consumed most of the lighter element through fusion, less energy is released, and the force of gravity takes over. The interior of the star compresses, pressure and temperature rise, and the fusion of heavier elements begins. As each element undergoes fusion, fewer and fewer consumable products are created.[17]

A rapid acceleration of burning takes place: helium proceeds to fuse, taking only 700,000 years; Carbon burning lasts for 600 years; Neon burning lasts for one year; oxygen burning about six months; and finally, silicon is fused to form iron in less than a day. At the center of the star, temperatures have reached three billion degrees Celsius. In reality, stars don't entirely burn an element, finish, and then go on to another. Instead, there is multiple simultaneous burning going on. As the star produces heavier elements toward the end of its life, they tend to gravitate toward the center of the star, a place that is hotter and capable of igniting their fusion reaction. The multiple fusion burning occurs in different layers, resembling the sketch below (stellar onion diagram). At the end of this dynamic process, iron accumulates in the core. The last day of the star's life is where the story of life begins on Earth, but before that happens, one of the most dramatic events in the galaxy will take place.[18]

Iron cannot undergo fusion to create heavier elements unless energy is added to the reaction. If iron fuses, it will cause the core of the star to cool. In this case, hydrostatic equilibrium is no longer possible. The tremendous weight of the star can no longer be supported: there is an implosion. The iron core is about 1000 kilometers in diameter, and in one-tenth of a second, it collapses down to a radius of about 15 kilometers. As the iron nuclei race toward the center of the dying star, nuclei are crammed up against nuclei. Electrons combine with protons to form neutrons, and a particular particle, the neutrino, is released in the process.[19] A compact object called a neutron star is left where the core of the giant star once was.

What happens in this brief moment leads to a brilliant event: a supernova. Most of the energy of the implosion is carried away not by visible light, but by the elusive neutrino. The neutrino subsequently creates a shock wave of photons as they race out of the star. It takes around two hours for the shock to reach the surface of the star, but when it does, the star explodes, sending most of its material outward. It's a brilliant explosion; the star shines over 100 million times brighter than it did just hours earlier. The star then ejects leftover hydrogen, helium, carbon, oxygen, and nitrogen into space. Even metals heavier than iron, those which normal fusion reactions cannot create, are produced. The expanding shock wave carries with it newly produced silver, gold, lead, and uranium. This "star stuff," as astronomer Carl Sagan would call it, is sent to interstellar space to be one day reformed. The atoms that make our Sun, the Earth, and you and me all began here in a supernova: we are indeed made from the ashes of stars.

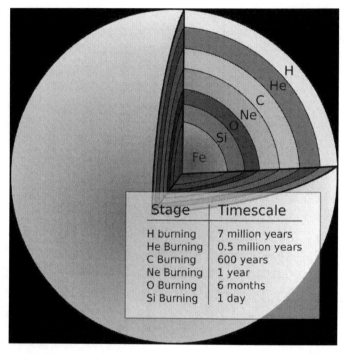

Sketch of a 25-solar-mass-star nearing the end of its life (Source: A.C. Phillips)

The most powerful supernova ever recorded shone nearly 600 billion times brighter than our Sun. When it exploded, it was 20 times brighter than all the stars in the Milky Way Galaxy combined.[20] A star releases ten times more energy in this explosion than all the energy our Sun does in 10 billion years. Had this supernova been placed at 8.6 light-years, the distance of the brightest star in our sky, Sirius, it would have shone like the midday sun. Neighboring planets to the supernova would have been ablated away to nothing—vaporized in the explosion.

If you live in the northern hemisphere around the midlatitudes and look up to the sky on a clear winter evening, you may see three brilliant stars in a row: this is the belt of the constellation Orion. Halfway from the belt and due north toward the bright star, Capella, is the Crab Nebula. The figure below shows a Hubble telescope image of the nebula.

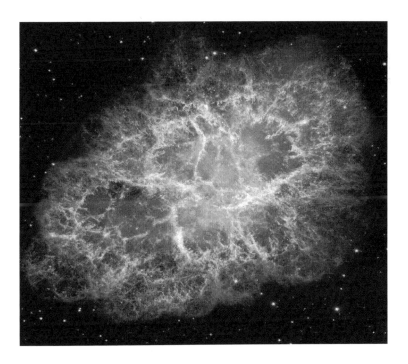

The Crab Nebula, a supernova remnant (Source: NASA and STScI)

The Crab Nebula is the remnant of a supernova explosion that occurred in the year 1054; Chinese astronomers observed the event. There is no chance for a planet around one of these stars to have had enough time to harbor life. But these stars are essential to us. Without them, most of the periodic table would not be populated. Later generations of stars need the by-products of these giants for life to be possible on one of their orbiting planets. Supernovae are the violent start of our existence and where our story begins.

Eventually, through the gentle work of gravity, over eons of time, the matter ejected from a supernova will reform into other stars and planets. Most of the ejected material from the giant is unused hydrogen and helium, and that will eventually be used as fuel for the next generation of stars.

The Pleiades (Source: NASA, ESA, AURA/Caltech, Palomar Observatory)

We can actually see gas and dust reforming to produce new stars. On that same clear winter night, again look up to the belt of Orion. If you look straight at the belt and then draw your eyes directly toward Polaris, the North Star, about the distance of five lengths of the belt, you will see a fuzzy object in the sky. This cloudy object is the Pleiades, also known as the Seven Sisters, shown in the figure. This open star cluster is a group of 3000 relatively young stars, but we can observe the seven brightest without the aid of a telescope. These stars formed out of the interstellar gases left from previous stars. Though the Pleiades are beautiful to watch from a very far distance, these stars radiate ultraviolet light that would kill any life on a nearby planet.

Planetary astronomers are relatively confident that about 80 % of all the stars have some planetary system orbiting them.[21] But just because a star has a planet, there is no guarantee that one of these planets is habitable. As we have seen, that star must be in the right location, and of course, the planet must have the right material. But there are other factors as well. Let us start with our star, the Sun, and see what makes it unique.

Our special star: The Sun

Our star, the Sun, has the right characteristics that make it almost ideal for sustaining life on a nearby planet. The Sun is virtually perfect in supplying energy to the Earth, not only in the past but today and for billions of years to come. The energy it supplies is sufficiently powerful to warm our planet that is tens of millions of kilometers away. Our Sun is a massive collection of mostly hydrogen and helium gas held together by the powerful forces of gravity.[22] Gravitational attraction is so strong near the center of the Sun that ordinary hydrogen atoms are ripped apart. The single proton, orbited by a single electron, is torn apart so that free protons and electrons coexist in a sort of soup of particles (a *plasma*). Protons are so densely packed that there is a high chance they collide with other protons. As previously described, protons fuse to form heavier elements, giving off gamma rays and X-rays in the process.[23] This powerful form of radiation starts its long journey toward the surface of the Sun.

Usually, light would take just a second to cover the distance from the Sun's core to its surface, but because of the number of hydrogen and helium nuclei in the way, it will take that individual photon up to 100,000 years to exit the Sun.[24] As the light "bounces" off these other particles, it loses energy, not by slowing down, but by broadening, so to speak, in its frequency.[25]

High-energy X-rays and gamma rays become low-energy and, therefore, low-frequency radio waves, although some very high-energy light rays also make their way to the Sun's surface. On Earth, we receive the entire bath of all types of radiation, and occasionally, during a solar flare event, as shown in the figure below, additional radiation from the Sun is emitted and eventually reaches the Earth.

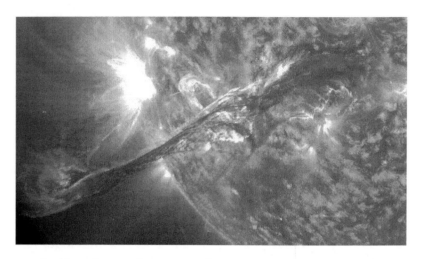

Our local star, the Sun, and a solar flare (Source: NASA Goddard Space flight Center)

Though our Sun is critical for life to exist on the Earth, it's nonetheless a raging nuclear furnace. The same energy production found in the destructive power of a hydrogen bomb is found here but on an astronomically larger scale. The Sun produces the equivalent of 100 billion one-megaton nuclear bombs every second, and this has been going on, in an almost constant fashion, for six billion years. As extreme as this seems, this is a fortunate situation, since the entire evolutionary process for life requires this uninterrupted energy. Any interruption of the Sun's energy production or impediments for that energy to reach the Earth would mean risk of the extinction of life on our planet.[26]

Our Sun is responsible for almost all the energy we use on Earth.[27] It's obvious where solar energy comes from, but wind energy also has its origins in the movement of air that results, in large part, by unequal heating of the Earth's surface by the Sun. All fuels from plants ultimately have their origins from sunlight energy as well. Even fossil fuels such as oil, natural gas, and coal, all originate from dead plants and animals, all of which obtained their energy from the Sun long ago. Without the Sun, our planet Earth would be a cold and miserable frozen rock with temperatures hovering around -400 °F (-240 °C).

In terms of its size and temperature, our star is considered "average." Many stars are smaller; they started off from smaller clouds of hydrogen gas. Less gravity from these stars means less compaction of hydrogen in their interior and the burning process is slower: the star is cooler. Stars that formed from more significant clouds of hydrogen gas form more massive stars that have stronger gravity. Hydrogen consumption in these stars is faster, and of course, these giant stars are hotter, more luminous, and shorter-lived. Astronomers have figured out that there is a relationship between the luminosity of a star and its temperature. Astronomers map this relationship in a chart called the Herztzsprung-Russell (HR) diagram, shown in the figure. Any star may be inserted in this chart, and our sun fits in somewhere near the middle.

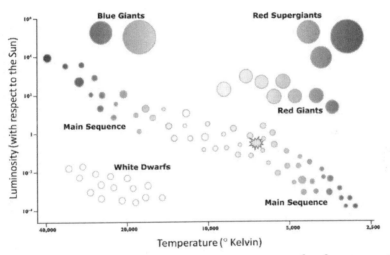

The universal luminosity versus temperature plot for stars—the Hertzsprung-Russell diagram. The Sun's location is also shown. (Source: Adapted from the Las Cumbras Observatory archive)

The surface temperature of a star ranges anywhere from about 3000 degrees Kelvin to about 25,000 degrees Kelvin.[28] We find hotter stars in the left part of the diagram, cooler stars on the right. The chart displays more luminous stars near the top of the diagram, and dimmer ones toward the bottom.[29] The band of stars running diagonally across the chart is known as the *main sequence,*[30] and here most stars reside.

It's very revealing just how few types of stars are appropriate for a habitable planet. The hottest, most brilliant stars appear in the upper left of the diagram along the main sequence. After their short lives, these stars expand as they burn heavier elements: they become supergiants, found in the top right of the diagram.[31] Planets around these stars have no chance for life to develop.

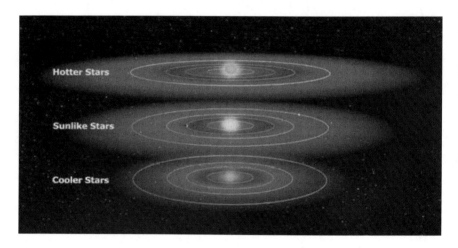

The CHZ for three different star types (Source: NASA Kepler Mission)

What about smaller stars? Could life flourish on planets around them? The smallest and considerably less luminous K-type and M-type (dwarf) stars are not much larger than our largest planet, Jupiter. The surface temperatures of these dwarf stars are so low that a planet such as our earth, orbiting at a distance that we currently assume, could not possibly support life: the planet would be too cold. Not only is there is a region called the galactic habitable zone, but there is also a region called a *circumstellar habitable zone* (CHZ).[32] Astrobiologists define the CHZ as the region around a star where water could exist on the surface of an Earth-like planet that orbits it. The CHZ changes for each star. The smaller the star, the less heat it gives, and the smaller and tighter its CHZ, and vice versa for a more massive star.³³ The figure shows the variation in the size of the CHZ and the relatively narrow band for our solar system (middle example).

The smaller CHZ around these dwarf stars creates other problems for planets: tidal forces. Though gravity decreases for less massive stars, tidal forces do not decrease the same way. Planets orbiting K- and M-type stars cannot be too close to their heat source, or they will be ripped apart by tidal forces. These same forces will cause continual squeezing on a planetary scale and result in large-scale volcanism.³⁴ Larger tidal forces can even cause planets to break up at their onset, soon after or while they are forming.³⁵ Smaller planets tend to hold together better than larger planets under tidal forces, but if a planet is too small, it will not have enough gravity to retain an atmosphere. There are strong indications that the smallest of stars, those to the right of the Sun on the main sequence, cannot have planets that are habitable due to the number of constraints that come with their small size.³⁶

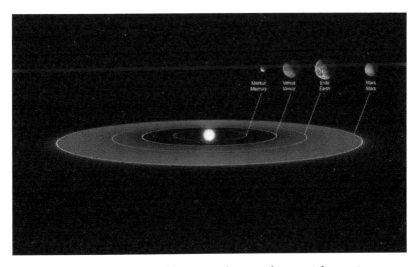

The circumstellar habitable zone shown along with our inner solar system (Source: ESO/M. Kornmesser)

Planets that orbit larger and more massive stars, those to the left of the Sun on the main sequence, have a different problem. The large gravitational field produced by these stars creates increased pressures in the stars' interiors. This pressure causes nuclear fusion to speed up, resulting in more energy released, more luminosity, and shorter lifetimes. Stars even slightly larger than our Sun, for example, between 1 and 1.5 solar masses, burn through their hydrogen fast enough that a potentially life-conducive planet will not have a reasonable amount of time for life to appear. A star of 1.2 solar masses, for example, has a lifetime of only 3 billion years. Scientists believe that life on planet Earth began only about 3.8 billion years ago, or about 2.2 billion years after the earth was created. It was only about 530 million years ago that centipede-like creatures began to explore the world above the water. Humankind would not appear until more than half a billion years later. Though massive stars shine magnificently and create elements needed for life for future-generation systems, they are, for the most part, not suitable places for a habitable planet. Our Sun, a "yellow dwarf star," is a perfect size and temperature needed for a habitable planet.

But it's not only the type of star that is important for a viable habitable planet; timing is also essential. Our sun formed from a gas cloud, someplace in the arms of the Milky Way Galaxy.[37] At that time, the young Earth was an utterly inhospitable place for life. Six billion years ago, our sun emitted dangerously high-energy radiation. These emissions would have sterilized any life on earth at that time.

But our star is in its midlife. Fortunately, a star the size of our sun undergoes no midlife crisis: as we have said, it will go on burning hydrogen for about another six billion years just as it has been doing up to now. Long before the Sun existed, other stars spewed hydrogen, helium, and some heavier elements into the interstellar medium. Gravity brought everything together, and our solar system was created. For millions of years, the Sun was not behaving well. Even if earth could have supported life otherwise, the Sun would have prevented it. Young stars similar to ours have been found to blow off their outer layers of gas into space, creating a dangerous environment for anything around them.[38] After about 100 million years of contracting, fusion ignited in its core, and the Sun soon settled down and became a "boring" adult star that could provide steady light to planet Earth for a very long time.

But our star will not live forever, and when the Sun nears the end of its life, it will undergo another transformation, a dramatic one. Hydrogen in the core will be mostly exhausted, resulting in gravitational contradiction and subsequent heating. The next heavier element, helium, will itself fuse to become carbon. This process will continue for a relatively brief, few hundred million years. The hotter and more compact core will cause the Sun's atmosphere to push out and cool. If some future civilization were present in those days, they would witness a star very different from the one we see today. The Sun's tenuous atmosphere would only be about half as hot as it is now, about 2,200 to 3,200 degrees Kelvin as opposed to today's temperature of 5,778 Kelvin. This cooler temperature would cause the Sun to appear red; in other words, the Sun will have entered its red giant phase.

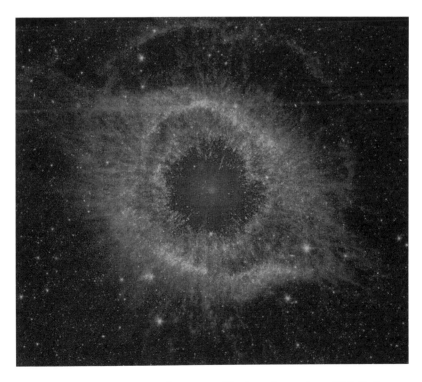

The Helix Nebula in the constellation Aquarius, 700 light-years away (Source: NASA/JPL-Caltech/Univ. of Ariz)

Though the red giant is somewhat cooler, it will not offer a pleasantly cool habitat for the inner solar system planets. The giant will grow to engulf the innermost planets Mercury and Venus, and perhaps even the Earth itself. All life on our planet will have long disappeared. The tenuous atmosphere of the red giant will reach our planet, evaporating all the oceans and leaving the ground as lifeless hot dirt.

Eventually, after a few hundred million years as a red giant, the Sun will have burned most of the helium in its core. The core will cool, shrink, and compress. More helium will eventually reach the core and ignite in a powerful fusion reaction that will blow the outer layers of the dying star into space, and into the interstellar medium. The gas and dust thrown out will provide material for a new star and new solar system sometime in the distant future. The exploded star will become one of the most stunning natural events in space, a planetary nebula. There are many examples of planetary nebulae, but one of the most beautiful is the Helix Nebula, shown above.

Our Sun will end its long life much like the Helix Nebula. The star that remains will become a compact *white dwarf* with a mass like that of the Sun but about the same size as the Earth.[39] No nuclear fusion reaction occurs at this point, but residual heat will slowly radiate away over billions of years. Although a white dwarf offers a very stable and constant heat source for planets, any planet remaining after the red giant phase will have suffered irreparable destruction. We expect no planets to support life around a white dwarf star. In summary, though our earth is currently a virtual paradise for life, it will one day die in the throes of a planetary nebular explosion.

The massive star that produced a supernova and the subsequent Crab Nebula came from a star that was about 9 to 11 times more massive than our Sun.[40] As inhospitable as our solar system will be when our sun reaches its final days, the space surrounding a supernova will be worse. Radiation from a supernova will sterilize planets as far away as 100 light-years from the explosion. The star that is left behind after a supernova explosion makes a white dwarf star look tenuous. The end phase of a star with a solar mass of 9 to 11 solar masses is called a neutron star. These types of stars are essentially one large "atom nucleus," held together not by nuclear forces, but by gravity.[41]

This star, once as large as our inner solar system, is crushed down to the size of a small city. A single teaspoon of neutron star weighs as much as the entire Rocky Mountain Range.

As inhospitable as the situation is near a neutron star, an even more bizarre fate is left to a star that is more than about 20 times the mass of the Sun. At this point, there will be nothing to keep this type of star from imploding to a *singularity*, an object with no dimension. Black holes were theorized from Einstein's field equations (1915), but their discovery had to wait until 1971, and the first observations of the region surrounding a black hole didn't occur until early 2019. Scientists call these objects black holes because they are black: no light escapes them. Nothing can stop the implosion of the core of a sufficiently massive star. Gravity, generally the weakest force, now becomes the dominant force, overriding even the strong nuclear force. In a neutron star, neutrons are packed next to each other, preventing further collapse, but once critical mass is reached, nothing in nature can override the force of gravity.[42] The region surrounding the singularity would look totally black to an observer. Also, space surrounding the black hole is greatly distorted due to the curvature of space caused by the intense gravity. Not many decades ago, black holes were considered things of the imagination. Now it's understood that black holes are found in many places in our galaxy and in other galaxies.

Our Milky Way Galaxy is a dangerous place, with an estimated 100 million neutron stars and 100 million black holes. Many other objects render large regions of space in our galaxy utterly incompatible for a life-bearing planet.[43] But fortunately for us, the closest neutron star is 250 light-years away, and the nearest black hole is 1600 light-years away.

In summary, of all the different types of stars, only a tiny fraction of them are capable of hosting a planet with life. And among these stars, only a smaller fraction are in the right place in the galaxy at the right time. But even if a planet forms around a proper G-type star, far from the chaotic core of the galaxy and the dangerous galactic arms, other hostilities are out there. If one is to consider life on a planet resulting from pure chance, then we must also consider the dangers in a planet's backyard.

The special backyard: Our solar system

We live in a special solar system. The planets that make up our solar system are very different from one another. In terms of life, those differences make a planet either heaven or hell. A planet must also be the right size, not too large, nor too small. One that is too small will not have enough gravity to hold an atmosphere, and without an atmosphere, there is not much chance for organisms on land to survive. Of course, aquatic life can still exist, but without an atmosphere, temperatures on that planet will vary widely between night and day. Water, even if it could exist on a planet without an atmosphere, would either evaporate or be frozen solid. The small world of Mercury, for example, suffers this fate. During a Mercury day, temperatures reach 800° F (427 °C); at night, temperatures plunge to -290 °F (-180 °C).

Mercury is the closest planet to the Sun, orbiting at a distance between 28 and 43 million miles (46 and 70 million kilometers). We have seen the devastating consequences for planets that are too close to their star: total destruction. A less dramatic outcome, but still serious, is tidal locking: the orbit and rotation of the planet can become synchronized. In other words, one side of a planet will forever face its sun; the other will be in perpetual nighttime. Mercury is not quite in a tidally locked state, but "the planet spins very slowly—three times every two Mercury years, or once every 60 Earth days."[44] The extremely long Mercurian days scorch the surface for weeks; this is followed by a brief period when temperatures pass through mild Earth-like temperature ranges at sunset, before the planet's surface is plunged into weeks of unbelievable cold. The harsh conditions found on this innermost planet provide no chance for life.

In terms of harboring life, giant planets also have problems. We know this because we have four such planets in our solar system: Jupiter, Saturn, Uranus, and Neptune. Through technology, we have been able to observe these giants up close. The closest and largest is Jupiter. Its enormous mass creates such a strong gravity field that it retains the lightest elements in its atmosphere. Ninety percent of Jupiter's atmosphere is molecular hydrogen, and about 10 percent is helium, a percentage similar to that of our Sun.[45] Also found in Jupiter's atmosphere are traces of hydrogen sulfides, methane, ammonia, water vapor, and even oxygen. Strong windstorms reaching over 600 kilometers per hour move these clouds of gases around. When sunlight strikes their upper atmospheres, giant swirls are created, as illustrated by the picture taken by NASA's Juno probe.

Though detectors have found some water in Jupiter's atmosphere, it's an unlikely candidate to support life. First, there is no distinct bottom to Jupiter's atmosphere—no land or oceans. The atmosphere gets thicker and thicker, building up to a nearly incompressible liquid. Gaseous hydrogen gives way to liquid hydrogen, and then as the pressure increases, hydrogen transforms into a special state called liquid metallic hydrogen. Under extreme pressure, hydrogen takes on properties of a metal. The rapidly spinning planet (one rotation just under 10 hours) causes liquid metallic hydrogen to produce a strong magnetic field, 20,000 times stronger than that found on Earth. Jupiter's magnetic field is not deadly per se, but charged particles from its *solar winds*[46] get trapped in the bands of its magnetic field—and these particles are deadly. The upper atmosphere of Jupiter stops this harmful radiation, but the inner moons of Jupiter, Io and Europa, are frequently exposed to sterilizing radiation.

Mercury's geology: false color image taken by the Messenger spacecraft (Source: NASA/JHUAPL/CIW

The Jovian[47] environment is harsh. Some have speculated that floating microbes could survive somewhere in the mid-atmosphere of Jupiter. Perhaps the combination of energy from lightning flashes, chemicals for life, and heat generated by the planet could, at least in principle, support microscopic life. But even if life were somewhere in Jupiter's atmosphere, it would have to maintain its altitude. Climbing too high would expose any life to radiation or extreme cold; falling too deep in the atmosphere would cook any living creature by the heat generated from the planet's interior. Most likely, Jupiter does not support life, and for similar reasons, none of the other giant planets are considered potential hosts for life.[48] For life to be found on a planet, especially higher forms of life, a planet has to have many things going in the right direction; it must have the appropriate mass, be at the right location, and be orbiting the correct type of star. But there is more.

The Earth | 99

NASA's Juno space probe picture of the giant planet as it flew just over 11,747 miles above its surface (Source: NASA/JPL— Juno mission)

It's not just our planet or the other planets that make our solar system unique; the arrangement of the planets is critical. An intruder can alter a peaceful, habitable planet's global environment and turn it into a "living hell." A comet is a small body of rock and ice that orbits the Sun. When it approaches, it warms and releases water, dust, and gases such as nitrogen, carbon dioxide, ammonia, hydrogen, and methane. Comets are not big, ranging in size from only 100–200 meters (330–660 ft) and occasionally found to be as large as 30 kilometers (19 miles) across.[49] These tiny bodies are thought to number in the billions; they are located mostly outside the orbit of Neptune in the expansive cloud of comets, the *Oort Cloud*. When our sun passes close to another star, the gravitational bump from the passing star may cause comets to head toward the Sun in a highly eccentric orbit. As the comet approaches the Sun, a *coma*, or a diffuse cloud of gas and dust surrounding the nucleus of a comet, is produced.

This nebulous envelope around the comet can reach a size comparable to the full moon.[50]

Though a comet is beautiful when traveling across the sky, they can be very destructive if one hits the Earth. On the morning of June 30, 1908, in Tunguska, Siberia, a remote place in Russia about 800 kilometers north-northwest of the Great Lake of Baikal, an object about 60 to 190 meters (200 to 620 feet) was thought to have entered into our upper atmosphere. The object exploded half a mile to a mile above the ground. The explosion, known today as the Tunguska event, devastated a forest area of 829 square miles (2150 square kilometers). Some 30 million trees were flattened or scorched. Londoners, living thousand of miles away, heard the pressure wave as it traveled across the northern hemisphere. The figure below shows the ground-zero area that was recorded on film some 20 years later; the figure also shows a recent aerial picture for comparison. The most reasonable theory to explain the Tunguska event was that a small chunk of comet broke off and entered a collision course with earth. The size of the object, though large for a meteorite, was still small enough that it broke up in the atmosphere. The loosely compacted material that forms a comet would have allowed for more natural disintegration of the incoming projectile.

The 1908 Tunguska event, 20 years later (left) and 100 years later (right) (Source: Science/NASA, Physics.org, and Science.com)

The explosion in this remote spot in Siberia represented one of the largest in modern history. Depending on the height at which the burst took place, the energy relased is estimated to be in the range of 3 to 30 megatons of TNT. The midrange estimate of 15 megatons represents an explosion 1000 times that of the atomic bomb dropped on Hiroshima, Japan in 1945. The blast resulted from only a small fragment of comet or asteroid. What would have occurred if a more substantial piece had entered the atmosphere?

The famous Halley's Comet has a mean diameter of 11 kilometers.[51] If this comet crashed to the Earth, the devastation would be global. Fortunately, events of this size do not happen often, but an event of this scale might have occurred around 65 million years ago when a comet or asteroid collided with our planet. A partial circle, 180 kilometers wide, can be found on the northwest corner of Mexico's Yucatan Peninsula in the Chicxulub crater. Geologists and planetary scientists believe this crater to be the remnant of one formed when a massive object, 10 kilometers across, impacted the ground. The destructive force was almost unimaginable, creating a gigantic wave some 100 meters high that raced across the Gulf of Mexico and sent debris as far as New Jersey.[52] A cloud of superheated dust, ashes, and steam rose and spread throughout the Earth. The dust remained in the atmosphere, perhaps for decades, causing sunlight to be blocked and subsequently cooling the planet.[53] What would happen if a similar event occurred today?

In 1995 we witnessed an event that brought us close to the drama of Chicxulub. Two years earlier, Carole and Eugene Shoemaker and David Levy spotted a comet for the first time.[54] The comet they discovered was an unusual one. There were no single comet nuclei, but instead, a long cluster of nuclei was strewn across space. Further observations revealed that the comet was orbiting Jupiter, not the Sun. About 20 to 30 years earlier, the giant planet had captured this comet in its orbit. By detailed backtracking, astronomers determined that the comet passed close to Jupiter in 1992, entering the planet's Roche limit and causing tidal forces to break up the comet into smaller fragments. The larger fragments ranged up to two kilometers in size.

The astronomical community was fixated on this comet until the object came to a dramatic finish two years later. On July 16 and July 22, 1994, and at a speed of approximately 60 km/s (37 mi/s), the comet fragments collided with Jupiter's southern hemisphere. Multiple impact marks were made on Jupiter, and the larger ones were bigger than the Earth.[55] If this comet or even a piece of it had hit the Earth, the destruction would have had worldwide consequences. The dust cloud created by the impact, though less intense than the Chicxulub cloud, would have covered the planet for years, shutting out the Sun and creating an artificial winter around the globe. Most likely, an impact on earth would happen in the oceans, but ironically, this would not be a blessing. Had a comet like Shoemaker-Levy hit the earth in midocean, the wave generated would have been miles high and would have ripped through every coastal area on the planet. In some places, the wave would have reached hundreds of miles inland—changing our world forever.

We are very fortunate that events like the Shoemaker-Levy comet don't happen often. The spectacular phenomenon of July 1994 demonstrated the reason: our giant neighbors have been protecting us all along. All four Jovian planets orbit the Sun more or less in the same orbital plane as the Earth. Comets are drawn to the massive giants, not to us, when they enter the solar system. These giants' large gravitational pulls have saved us from countless comets throughout the six billion years of our planet's existence. This protection is essential for life to start, flourish, and evolve. But the story does not end here.

Our home, the Earth

We live on a remarkable planet, exceptional in many ways. If we let our imagination run free, we could conjure up extraordinary places—beautiful, sublime beaches of the South Pacific, the mysterious Central African rainforests, an idyllic lake in the Swiss Alps. These examples of our blue planet are just a few in a place that is unique in the universe, at least as far as we know today. Maybe someday, we'll find another one like it. A total of 4260 *exoplanets*, or planets orbiting stars outside our solar system, have been found. This does not mean that we can peek into a telescope and discover an exoplanet with a blue ocean, forest, and beaches; it only means some planets could potentially harbor life.

Our closest neighbor, the moon, is a lifeless rock, void of water, at least on its surface, and absent of any atmosphere. This stark, dirt-strewn satellite may seem like it serves little purpose except, for example, having been the target destination for space missions in the Cold War era. And we all admire a full moon on a clear night as it lights up the evening. But our large satellite plays a role in our planet's habitability, in a way that may not seem obvious. To begin with, let's consider the Earth's seasons.

The earth is tilted 23.5 degrees on its axis;[56] this tilt is crucial for Earth's habitability. If our planet rotated without a tilt, there would be no seasons, and as a result, the earth would have stratified climate bands. Unsurprisingly, the further you traveled north, the colder it would get, but the temperatures would never vary. The Arctic and Antarctic regions would be forever frozen. The tropics would be as they currently are. Though life could have evolved on a planet without a tilt, humans could never survive the endless winter at high latitudes. According to Don Attwood, an ecological anthropologist at McGill University in Montreal, without seasons,

> ...humans would probably never have advanced past a state of living in small, scattered settlements, scrounging for survival and often dying of horrific insect-borne diseases.... The amenities of modern civilization cannot be built on such a foundation.[57]

Others speculate that drastic climate changes due to a changing tilt would allow for only "small, robust organisms to survive."[58]

Fortunately, the earth does have a tilt, and that tilt cannot easily change its axis of rotation. The moon locks us into the 23.5-degree tilt from our orbital plane, and this provides for regular and needed seasons. We can spin a toy top on a table and give it an initial tilt. The toy top remains in this tilt though it rotates (precesses) around the vertical axis. Similarly, the earth spins on its north-south pole axis. This axis is tilted with respect to the earth's orbital plane. The moon keeps the earth's tilt steady, and a slow precession, a phenomenon called the *precession of the equinoxes*, takes 26,000 years to complete. Mars, for example, does not have a large moon, only two small satellites: Phobos, with a diameter of 22.2 km (13.8 mi), and Deimos, with a diameter of 12.6 km (7.8 mi). These two moons are not large enough to keep the red planet from wobbling; its polar ice caps were once at its equator. More than a beautiful object, the moon serves a profound purpose for the Earth.

Closer to home, our atmosphere shields us; it acts as a protective shroud for all life below. The topmost layer, the *ionosphere*, reaches to about 250 to 1000 km (150 to 620 miles); at this height, gases interact with *extreme ultraviolet* radiation. A neutral atom absorbs

The northern lights created by harmful charged particles from the Sun (Source: Scott Kelly/NASA)

incoming radiation and loses one or more of its electrons, becoming ionized and giving rise to its name. Without this protection, radiation and high-energy particles from the Sun and deep space would race toward the ground and rip through any living tissue. We are aware of the power of high-energy particles as they interact in the atmosphere by the light that is given off. When the ionized elements recombine with an electron, a photon is produced. These events occur near the Earth's poles almost every day; this is where we find beautiful *aurora* displays, as seen in the figure above.

Auroras remind us of another beneficial characteristic—the Earth's magnetic field. A display of lights is produced when the magnetic field acts on charged particles. The core is made mostly of iron; as the Earth rotates, that iron generates a strong magnetic field. The magnetic field deflects highly energized particles and directs them into "particle traffic lanes" toward the poles, where they enter safely into the upper atmosphere.

Next, the stratosphere is found at an altitude of 10 to 50 kilometers (6 to 30 miles), and a significant component of the stratosphere is the protective *ozone* layer. Ozone (composed of three oxygen atoms) is very rare, averaging only about one molecule in every 10 million air molecules. But despite it being rare, ozone plays a vital role in absorbing dangerous ultraviolet light coming from the Sun. Many clinical studies have shown that exposure to ultraviolet radiation is harmful to humans, and without the ozone layer, we would be irradiated daily by the Sun.

Below the stratosphere, we arrive in the *troposphere*, a layer from ground level to about 3.7 to 6.2 miles (6 to 10 kilometers). About 75 % of the atmosphere's mass is located in the troposphere, and this is where nearly all weather takes place; it's also where life abides. Our atmosphere is composed of a fragile balance of elements: 78 % nitrogen; 21 % oxygen; and almost 1 % argon with a trace amount of other gases, including carbon dioxide (0.04 %). We depend on oxygen, and 21 % has been determined to be just about the right and safe amount.[59]

No simple formula keeps the concentration levels of gases in the atmosphere at the right levels. Appropriate levels for life are maintained via a complex interplay of physics, chemistry, and biology.[60] While all this is happening, there is also an exchange of gases between the atmosphere, the ocean, and the ground. The earth is one huge dynamic and complex set of systems, analogous to a living being, where substance is continuously passed from one part to another for its well-being.

Given the complexity of the atmosphere and our dependence on it, what would happen if that balance was disrupted? Would the Earth's atmosphere naturally correct itself back to its proper state? In the remote past, interactions between the *biosphere* and the Earth have produced *runaway feedbacks*; this was the case in the *Great Oxidation Event* that occurred 2.3 billion years ago.[61] The atmosphere completely changed composition in a relatively short amount of time. When these rapid changes occur, they are often associated with dramatic and bad consequences, but not always. The Great Oxidation Event, though it saw dramatic changes in the atmosphere, was ultimately very beneficial for life, including higher forms of life. External crises can also play a role in changing the delicate system of balance on Earth. The disappearance of dinosaurs, the rise of mammals, and eventually, the appearance of man all came about, to a certain degree, by the effects of the massive comet mentioned earlier. But the question remains open: Can life impact from living creatures, including humans, lead to detrimental changes to the atmosphere and the planet? Without a doubt: yes.

Elevated carbon dioxide levels in the atmosphere are of concern, as this is related to the warming of the planet in recent decades. If this gas were too prevalent in the atmosphere, we would continue to heat up. In the extreme, our atmosphere would become Venusian-like. Our sister planet, very similar to Earth in many ways, has a carbon dioxide level of 96.5 % and surface temperatures of 864 degrees Fahrenheit (462 degrees Celsius). But too little carbon dioxide in the atmosphere is also harmful, since plants could not maintain sufficient levels of photosynthesis. Animals, as well as humans, would subsequently suffocate.

We can tell a similar story about water vapor. Too much water vapor in the atmosphere would result in a runaway greenhouse effect, but too little would result in the planet being too cold to support life.[62] The Earth may be a paradise, but it's a fragile one. Human beings can alter their environment by overexploiting resources and by using them in an unsustainable way. A rapid anthropogenic planetary change has been occurring for the past 150 years.[63] Currently, the Earth is on a damaging path in a rapidly changing climate phase through anthropic greenhouse gases. Though this problem is somewhat neutral to the discussion of faith and science, the entire situation only adds to the uniqueness of the story: the universe we know of has only one habitable planet, and that took billions of years to support us. As a species, we are singlehandedly changing this world, and all this in only a few decades.

Water is undoubtedly one of the essential molecules on Earth, found in every puddle, every stream, every lake and ocean, and in liquid form in every living organism.[64] If we expect to find life elsewhere on some planet, we would expect this simple chemical compound to be present. The combination of water's properties is exceptional. In its solid form of ice, it's lighter than its liquid form of water, an anomaly that causes it to float.[65] Had this not been the case, lakes and oceans would have ice at the bottom. The continual freezing of water by the atmosphere would mean that lakes and large parts of the oceans would freeze solid in winter, killing all forms of life. But since ice floats, it insulates the water below it, protecting fish and other animals. Floating ice keeps the planet 20 degrees warmer than it would be had the ice sunk to the bottom.

Since water is a small molecule, it's also very "fluid"; it's ideal for moving chemicals around, to and from an organism. At the same time, liquid water does not like to change phase easily, and this is a good thing. It takes a great deal of energy to cause liquid water to turn to gas[66] or to change from a liquid to a solid.[67] Water forms *hydrogen bonds*, a weak bond between the positive proton from hydrogen and the electronegative oxygen, as shown in the figure on the next page. Energy is required to break this bond before water can boil.[68] This bond causes water to remain in its liquid state for most of the range of temperatures our planet Earth offers. If this were not the case, lakes and oceans would freeze, and life would not be possible. Also, the high heat necessary to vaporize water means that less water vapor is present in our atmosphere than otherwise would be.

In summary, none of the details given here, from the smallest molecular bond to large-scale global dynamics, should be taken for granted. Life is intrinsically linked to its surroundings, and those surroundings are fragile.[69]

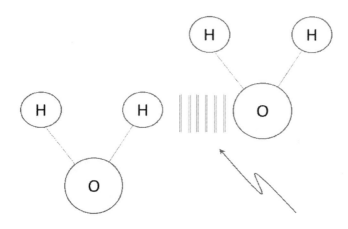

Hydrogen Bond

Sketch of water molecules including oxygen (O) and hydrogen (H), and a hydrogen bond between two molecules

What if?

How unique is the Earth? That question is foremost among *astrobiologists*, scientists who look for life outside our solar system. Significant progress has been made by science in the last few decades in search of an Earth-like planet. According to the best estimates in 2019, the observable universe contains an estimate of a billion trillion stars (1 followed by 18 zeros), and it's estimated that 10 to 30 percent of these stars have planets—that's plenty of planets, and the numbers may even be higher.

According to the work of Dr. Eric Zachrisson at Uppsala University in Sweden, since the Big Bang 13.8 billion years ago,[70] a staggering 7×10^{20} planets are thought to have been formed. Also, at the time of this writing, about 30 candidate exoplanets orbiting in the stars' planetary habitable zones have also been found; extrapolating this count leads to about 50 billion planets in the Milky Way alone.[71] Given this number, might we expect life to be teeming through the galaxy and cosmos?

It's easy to get swept away by exciting discussions of exoplanets and their discovery, but as we have seen in this chapter, many conditions have to come together for life to have a chance.[72] What is not conveyed in these numbers is the scarcity of exoplanets with Earth-like qualities that we know can support life, even higher levels of life.

The "right" characteristic	Generous odds	Number
Starting number		7×10^{20}
Type of galaxy	1:25	2.8×10^{19}
Place in the galaxy	1:1000	2.8×10^{16}
Kind of star	1:1000	2.8×10^{13}
Distance to star	1:50	5.6×10^{11}
Mass of planet	1:10	5.6×10^{10}
Spin of planet	1:5	1.1×10^{10}
Tilt of planet	1:10	1.1×10^{9}
With a substantial moon	1:10	1.1×10^{8}
Magnetic field	1:10	1.1×10^{7}
Planet not near any dangerous object * (e.g. black holes, nebulae)	1:50	2.2×10^{5}
Planet in solar system with a large Jovian-like planet for protection	1:50	4500

Statistics for "Earth-like" planets based on reasonable assumptions and generous odds (Source: Adopted from J. Clayton, See reference)

The constraints put on a planet, one resembling the Earth, substantially reduce the total number of viable planets—only about 4500 for the entire universe, significantly less than one Earth-like planet in our Milky Way Galaxy.[73] In other words, an Earth-like planet is extremely rare. Even if the odds were made incredibly favorable (substituting 1 in 10 for each value in the table), the result would be 7×10^9 Earth-like planets throughout the universe, or less than one for the Milky Way.

Dr. Zachrisson also suggests that most planets in the universe shouldn't look like Earth.

> [Zachrisson's] model indicates that Earth's existence presents a mild statistical anomaly in the multiplicity of planets. Most of the worlds predicted by his model exist in galaxies larger than the Milky Way and orbit stars with different compositions—an important factor in determining a planet's characteristics. His research indicates that, from a purely statistical standpoint, Earth perhaps shouldn't exist, and this is a "revelation that's both beautiful and terrifying at the same time."[74]

If the Earth is unique, what does this mean in terms of science and faith? If we cannot discover another planet that harbors life, our earth is positioned in an extraordinary place in the universe. Moreover, though it's impossible to prove that the Earth is unique, should not the observer at least consider the possibility of a miraculous intervention, independent of when that intervention came, either before time existed or during the unfolding of the universe's history?[75]

Turning the tables, what if another planet is found that harbors life? What does this say in terms of God? The answer is simple: nothing. Nothing precludes God from creating other life, including intelligent life, elsewhere.[76]

We still need to address one more major scientific issue. Up to this point, we have only explored the chance of finding an Earth-like planet in the universe, but the journey does not stop here. The very fact that you are reading this book, conscious, and yes, alive, is something to be considered. We open the door to what is likely to be the most exceptional single event since the Big Bang: the creation of life. Can unassisted nature create life, and what does that say about God?

Chapter summary and questions

We live in a special place in a special galaxy. Our Sun, solar system, and planet are all conducive to supporting life. The odds that another planet has all the beneficial characteristics we find here are exceptionally small. One researcher has estimated that our Earth is essentially unique in the universe—it should not even exist. Whatever the case, it's a miracle that we are here, and this adds credence to the supposition that an overseeing entity (a Creator) is responsible.

Questions to ponder:
- Recall that the anthropic principle is the principle that theories of the universe are constrained by the necessity to allow human existence. Does this affect your faith? If yes, how so?
- Though the anthropic principle is not testable (not falsifiable), it nonetheless makes a powerful statement about our place in the universe. How does this statement change your view of humanity?
- Questions for believers: What if extraterrestrial life was discovered; how would that change your view on God? What if that life was also intelligent? How would that discovery change your view? Why?
- Question for nonbelievers: What physical evidence, if any, would convince you that our position in the cosmos is unique? Do the multiple and independent characteristics of our Milky Way, our solar system, and the Earth that allow for life on Earth change your position? If not, why not?

NOTES
[1] David L. McMahan, *The Making of Buddhist Modernism* (New York: Oxford University Press, 2008).
[2] William Paley, *Natural Theology: or, Evidences of the Existence and Attributes of the Deity* 1st ed. (London: J. Faulde, 1802).
[3] Some argue that the universe is compelled to be as it is and therefore compelled to have conscience and sapient life emerge from it: this is the *strong anthropic principle* (SAP). Critics of this form of the anthropic principle (the author is one of them) have an issue with the universe being "compelled," unless of course, some nonrandom process was involved.
[4] This version is known as the *weak anthropic principle*. See R. Penrose, *The Emperor's New Mind: Concerning Computers, Minds, and the Laws of Physics*

(New York: Viking Penguin, 1990), Chapter 10. We note that the consideration of multiverses is important here—that discussion will come in Chapter 9.
[5] John Clayton, "The Soft Anthropic Principle," Program 6 of the series *Does God Exist*, http://doesgodexist.org, 2013.
[6] More precisely, the Local Group, a cluster of 54 galaxies that is about 10 million light-years across and is part of the larger Virgo Supercluster, a group of more than 100 galaxies that is about 110 million light-years across.
[7] Some estimates are as many as 400 billion stars.
[8] See figure at the beginning of Chapter 3.
[9] Supernova: The final death throes of a massive star, one that suddenly increases in brightness because of a catastrophic explosion and that ejects most of its mass into intersteller space.
[10] Black hole: The imploded remnant of a massive star creating a region of space with a gravitational field so intense that no matter or radiation can escape.
[11] Referenced as Messier object 16 or M16.
[12] Globular cluster: A large, compact spherical star cluster, typically of old stars in the outer regions of a galaxy. The Milky Way has somewhere between 125 and 200 globular clusters, each containing from 10,000 to 1,000,000 stars.
[13] Hubble Space Telescope survey is archived by the European Space Agency (2010), www.esa.int; astronomers have identified more spiral galaxies than elliptical galaxies, but this has to do with the fact that spirals are typically brighter than ellipticals. Ellipticals are expected to be more numerous than spirals. Elizabeth Howell, "What Are Elliptical Galaxies?" *Science & Astronomy* January 09, 2019, https://www.space.com/22395-elliptical-galaxies.html.
[14] Spiral galaxies are generally brighter than most elliptical galaxies, since the former has massive star formations found in the spiral arms, producing copious light. Ellipticals are composed of significantly older, smaller, and of course less brilliant stars. Therefore, heavy metal production is lacking in ellipticals: the "star stuff" needed for life is missing. There are other indications that giant elliptical galaxies may be the best place to find solar systems with life. Some argue that if a star with its habitable planet might have been formed under the right conditions in a spiral, but following a collision with an elliptical, that star, along with its solar system, might be transferred to the elliptical. J. Sokol, "Giant old galaxies, not Milky Ways, are best for life to thrive," *New Scientist* 1 August 2015, https://www.newscientist.com, last accessed April 2019.
[15] The sun's mass is 1.9×10^{30} kg.
[16] This balancing process is known as *hydrostatic equilibrium*. Two components of pressure are at work against gravity: Radiation pressure (photons released from a fusion reaction push other particles outward) and thermal pressure (particles hitting particles and pushing each other outward). These two pressures keep the crushing force of gravity in check.
[17] For example, it takes one nucleon (a proton) to make a hydrogen nucleus, 12 nucleons (6 protons and 6 neutrons) for carbon and 56 nucleons (26 protons and 30 neutrons) for iron.
[18] A. C. Phillips, *The Physics of Stars*, 2nd Edition (Hoboken, NJ: Wiley, 1999).

[19] The reaction is a proton and electron combining to form a neutron, emitting a neutrino: p + e- → n + v_e. The neutrino (specifically, the electron-neutrino, a "flavor" of neutrino that is associated with electrons) is emitted in the reaction. The neutrino is very unreactive with the surrounding star material, and most of the energy of the imploding star is carried off with it.
[20] Supernova ASASSN-15lh (All Sky Automated Survey for Super Novae) first detected on June 14, 2015, in its host galaxy 3.8 billion light-years away. Lee Billings, "Found: The Most Powerful Supernova Ever Seen," *Scientific American* (January 14, 2016), https://www.scientificamerican.com/article/found-the-most-powerful-supernova-ever-seen/, retrieved in August 2018.
[21] E. Siegel, "What Are The Odds of Finding Earth 2.0?" *Science* (May 10, 2016), https://sciencesprings.wordpress.com, last accessed April 2019.
[22] The Sun is mostly hydrogen (about 70 %) and helium (about 28 %) as well as carbon, nitrogen, and oxygen that make up 1.5 % of the Sun's mass. The other 0.5 % is made up of small amounts of many other elements such as neon, iron, silicon, magnesium, and sulfur.
[23] These "rays" represent very high frequency light, with wavelengths 100 to 1 million times shorter than visible light.
[24] The revolution of quantum physics in the twentieth century showed us that light can take on the properties of a wave and a particle (Chapter 3). We call light particles photons. Light from the core of the Sun bounces off not only other nuclei as it races toward the surface of the Sun, but also other photons.
[25] Frequency is strictly a wave property, not a particle property. This property describes how many crests of a wave go by per unit of time, or per second. A wave loses energy by stretching out, having fewer crests pass by per second (though its speed remains the same).
[26] On a few occasions, near extinction has happened (e.g., the Cambrian–Ordovician event 488 million years ago), but this was not a result of erratic behavior of the Sun.
[27] Only nuclear power, both fission and futuristic fusion, have no direct connection to the Sun. However, as we have already pointed out, uranium, a fuel used in the fission process, was originally created in a massive star long ago.
[28] Kelvin is the absolute scale of temperature and is often used in astronomy. The lowest temperature on this scale is absolute zero, 0 °K (Kelvin), or equivalent to -273 °C.
[29] Credit must be given here to women astronomers: lots of the tedious but important spectral classification of stars was done in the early twentieth century, and this was accomplished by women (and yes, they were paid less then men).
[30] The HR diagram represents a major scientific leap in astronomy. The stars in the main sequence are "middle-age" stars. Other stars, off the main sequence, are giants and dwarfs (stars that have reached the end of their lives). The HR tool, along with many others, allows science to make predictions and inferences about the part of nature we cannot physically see or measure.
[31] The largest being VY Canis Majoris (the "Big Dog Star"). The exact size is uncertain, but it has been estimated to have a radius between 1400 and 2100 times

that of the Sun. If placed where our sun is located, it would extend to the orbit of Jupiter or Saturn.

[32] Matthew Williams, "Could Habitable Zones Be a Lot Narrower Than We Thought?" July 4, 2019, https://interestingengineering.com/could-habitable-zones-be-a-lot-narrower-than-we-thought, retrieved in September 2019.

[33] S. Huang, "Occurrence of life in the universe," *American Scientist* 47:3 (1959): 397–402.

[34] Rory Barnes and René Heller, "Habitable Planets Around White and Brown Dwarfs: The Perils of a Cooling Primary," *Astrobiology* 13:3 (March 2013): 279–291.

[35] The *Roche limit* is that minimal distance in which a celestial body cannot be held together by its self-gravity due to this tug-of-war and the tidal forces acting on the planet. Édouard Roche: "La figure d'une masse fluide soumise à l'attraction d'un point éloigné" (The figure of a fluid mass subjected to the attraction of a distant point), part 1, *Académie des sciences de Montpellier: Mémoires de la section des sciences*, Volume 1 (1849), 243–262 (in French).

[36] M. H. Hart, "Habitable zones about main sequence stars," *Icarus* 37 (1979): 351–357.

[37] It's thought that most stars larger than the smallest and faintest M and K stars form in binary systems or systems of triplets or even higher multistar systems. Multiple star systems are not the places one expects to find life, as both gravitation and radiation fluctuations are not conducive to life on an orbiting planet.

[38] T Tauri stars are variable stars that fluctuate in brightness, named after the first one discovered in the constellation Taurus in 1952. They are newly formed stars, where hydrogen fusion has not yet started, and all their energy comes from heating by gravitational contraction. These young stars throw off material into the surrounding space; this lasts for about 100 million years. The TT in T Tauri has humorously been referred to as the "terrible twos" for a star.

[39] The details are technical, but suffice it to say, the unbelievably high densities of the white dwarf (1 ton per cubic centimeter) are kept from further collapse as a result of fundamental quantum mechanical properties.

[40] Gordon M. MacAlpine, et al., "A Spectroscopic Study of Nuclear Processing and the Production of Anomalously Strong Lines in the Crab Nebula," *The Astronomical Journal* 133:1 (November 30, 2007).

[41] In a neutron star, electrons are essentially absorbed by the protons and become neutrons. A neutron star is, therefore, one large mass of neutrons about ten kilometers across and with incredible density. Basic properties of quantum mechanics (the neutron degeneracy pressure or the resistance of neutrons being forced into the same space of another neutron) keep the neutron star from contracting even more.

[42] A singularity is a point in space (no dimension). Space and time are highly distorted around the singularity. No light that approaches within the region surrounding the black holes can escape. This region is called the event horizon, and it's typically small. After shedding its outer layer, a three-solar-mass black hole remaining after a supernova will have an event horizon of about 10 kilometers.

[43] The core of our galaxy is thought to have a supermassive black hole, about 4 billion solar masses. A. Boehle et al., "An Improved Distance and Mass Estimate for Sgr A* from a Multistar Orbit Analysis," *The Astrophysical Journal*, 830:1 (2016-07-19).
[44] Nola Taylor Redd, "How Hot Is Mercury?" *Science & Astronomy* (November 30, 2016) https://www.space.com/18645-mercury-temperature.html, retrieved August 2018.
[45] If a light molecule in the atmosphere is pointed upward and does not hit another molecule, it will continue on its way to outer space, never to return to the planet from which it came. The more massive the molecule, the slower its speed, causing it to "fall" back toward the planet. The gravitational fields of the massive Jovian planets force lighter elements such as hydrogen and helium to be retained.
[46] Solar winds are a stream of charged particles originating from the Sun and composed mostly of electrons, protons, and helium nuclei (alpha particles).
[47] The name Jupiter comes from the Roman king of the gods: Jupiter, or Jove, hence the adjective Jovian has come to mean anything associated with Jupiter or simply a Jupiter-like planet. This planet, along with its moons, was studied by Galileo Galilei in 1610 with the newly invented telescope (see Chapters 2 and 3).
[48] Moons of these planets, in particular, Saturn's moons Enceladus and Titian, might be candidates for microscopic life, perhaps found in a liquid layer of water beneath their icy shells, but we are a long way off from verifying this.
[49] Yanga R. Fernández, "The Nucleus of Comet Hale-Bopp (C/1995 O1): Size and Activity," *Earth, Moon, and Planets* 89:3 (2000).
[50] Pressure from the Sun's photons (*radiation pressure*) pushes the gases outward to produce a tail that may extend to about one million kilometers.
[51] "JPL Small-Body Database Browser: 1P/Halley" (11 January 1994 last obs), Jet Propulsion Laboratory, retrieved August 2018.
[52] Jonathan Amos, "Dinosaur asteroid hit 'worst possible place,'" BBC News, Science and Environment (15 May 2017), retrieved 19 August 2017, https://www.bbc.com/news/science-environment-39922998
[53] It has been postulated that the Chicxulub impact, roughly contemporaneous with the large-scale extinction of the dinosaurs, was connected with it. Paul Renne, "Time Scales of Critical Events Around the Cretaceous-Paleogene Boundary," *Science* 339:6120 (8 February 2013): 684–7, retrieved August 2017, https://science.sciencemag.org/content/339/6120/684
[54] Brian G. Marsden, "Eugene Shoemaker (1928–1997)," Jet Propulsion Laboratory (July 18, 1997), retrieved August 24, 2008.
[55] In contrast, the Chicxulub crater was formed by an object estimated to be between 11 to 81 kilometers (6.8 to 50.3 miles) in diameter. H. J. Durand-Manterola and G. Cordero-Tercero, "Assessments of the energy, mass and size of the Chicxulub Impactor," arXiv preprint arXiv:1403.6391, 2014.
[56] In other words, the plane that describes Earth's self-rotation is angled 23.5 degrees from the plane that describes Earth's orbit.
[57] Natalie Wolchover, "What If There Were No Seasons?" *Planet Earth* (March 9, 2012), https://www.livescience.com, retrieved April 6 2019.

[58] "If we had no moon," by David Foing, *Astrobiology Magazine* (Oct 29 2007), https://www.astrobio.net, retrieved April 6, 2019.

[59] If oxygen concentration goes below around 15 %, humans' ability to operate is increasingly impaired. If levels get below about 10 %, humans cannot survive for long. If the atmosphere had over 40 % oxygen, there would be a tendency for fires to start frequently. This is not to say that a robust species, and even an intelligent species, could not evolve on a significantly different planet with a much higher or lower oxygen content in its atmosphere.

[60] The biological process is part of the *biosphere*: the regions of the surface, atmosphere, and hydrosphere of the Earth (or analogous parts of other planets) occupied by living organisms. Their cumulative intake and output can greatly affect the atmosphere's composition.

[61] Photosynthesis from early plant life led to an abrupt rise in oxygen levels around 2.3 billion years ago. As a result, the concentration of oxygen in the Earth's atmosphere went from near-zero levels to the present level of 21 %.

[62] Other aspects of the planet contribute to a runaway change in climate. Warming temperatures, for example, melt the ice near the poles causing darker ground to be exposed. Dark ground absorbs heat better than ice and snow, which reflect much of the sunlight back into space. The Earth's reflectivity, or *albedo*, is the total amount of light reflected off the planet versus the total amount of light absorbed. A large change in albedo can contribute to runaway climate change. If the Earth's albedo were much greater than it currently is, we would experience runaway freezing; if it were significantly less, we would experience a runaway greenhouse effect.

[63] Carbon dioxide levels are at the center of the climate discussion, having changed from 0.027 % (280 parts per million) before the industrial revolution to 0.04 % (408 parts per million) in 2020. This seemingly small rise is significant enough to change the global climate.

[64] The water molecule, or H_2O, is composed of two hydrogens in a covalent bond with oxygen.

[65] The way water freezes causes the molecules to pack more densely than the average spacing of water molecules in a liquid. Some salts and metal alloys, as well as a few elements, arsenic, bismuth, gallium, germanium, and silicon, also expand when frozen, but none of these have the other life-essential properties of water.

[66] The heat required to do so is called the *heat of vaporization*.

[67] The heat required to do so is called the *heat of fusion*.

[68] Similarly, the melting temperature of water is considerably higher than other hydrides such as H_2S, with a melting point of -62 degrees C.

[69] Other Earth phenomena could be mentioned, including atmospheric transparency, earthquakes, volcanoes, and the spin of the Earth. All these unrelated activities contribute to the habitability of the planet. A good overview article can be found here: Charles Q. Choi, *What Makes Earth So Perfect for Life?* Live Science Contributor (October 18, 2012). Also, a classic work that is still relevant here is: James Lovelock, *Gaia: A new look at life on Earth* (Oxford: Oxford University Press, 1979).

[70] Shannon Hall explains Zackrisson's finding: though the number is extraordinary, most exoplanets are vastly different from Earth. Shannon Hall, "Exoplanet Census Suggests Earth Is Special after All," *Scientific American* (February 19, 2016).
[71] Erik A. Petigura, et al., "Prevalence of Earth-size planets orbiting Sun-like stars," *Proceedings of the National Academy of Sciences of the United States of America* 110 (31 October 2013): 19273–19278.
[72] For an interesting read on this topic, see Nathanial Scharping, "Why We Shouldn't Call Exoplanets 'Earth-like' Just Yet," *Discover* (October 21, 2018).
[73] More precisely, 4.5×10^{-8} Earth-like planets in the Milky Way, suggesting that the probability of finding one is essentially zero. Numbers are adopted from John Clayton's "The Soft Anthropic Principle," Does God Exist, Program 6, http://doesgodexist.org, 2013. (More realistic numbers, including place in galaxy 1:1000 and kind of star 1:1000, have been updated from earlier numbers of 1:100 and 1:25 respectively.)
[74] Nathaniel Scharping, "Earth May Be a 1 in-700-Quintillion Kind of Place," *Discover* (February 22, 2016), and Zackrisson et al., "Terrestrial planet across space and time," *The Astrophysical Journal*, 833:2 (2016).
[75] It's hard, if not impossible, to prove a negative, as illustrated in Bertrand Russell's famous "orbiting teapot analogy." Russel (an atheist) stated that the philosophic burden of proof lies upon a person making unfalsifiable claims. He would not expect anyone to believe that a teapot, too small to be seen by telescopes, orbits the Sun, though this assertion could not be proven wrong. Fritz Allhoff and Scott C. Lowe, "The Philosophical Case Against Literal Truth: Russell's Teapot," *Christmas – Philosophy for Everyone: Better Than a Lump of Coal* (Hoboken, NJ: Wiley, 2010), 65–66.
[76] More on this topic in Chapter 8.

Chapter 6 – First Life

At the dawn of the twentieth century, it was already clear that, chemically speaking, you and I are not much different from cans of soup. And yet we can do many complex and even fun things we do not usually see cans of soup doing. — Philip Nelson, Biological Physics: Energy, Information, Life

Nothing in nature comes close to the complexity of living creatures, and we don't have to look far to discover prodigious facts about life. Even at the base of the evolutionary tree, life is complex. For example, the microbacteria *Pelagibacter ubique*, or *P. ubique* for short, is impressive. Isolated as recently as 2002, this bacteria is among the smallest of self-replicating, free-living cells.[1] It measures about 0.5 micrometers, a hundred times smaller than the diameter of a human hair.

At the top of the evolutionary tree are mammals. The largest include the order of cetaceans—whales and dolphins. The 30-meter blue whale is the largest animal known to have ever existed, while the sperm whale has a brain size six times that of humans. Other cetaceans include dolphins and killer whales, which exhibit complex social behavior. But undoubtedly at the top of the tree is the order of primates, with the subspecies of homo sapiens (humans) at the very apex. The variety of living things goes beyond size; complexity is another issue. The way different parts of an organism function together is fascinating, whether that be different parts of the *P. ubique* or the various organs of a blue whale.

The variety of life: An Orange-lined Triggerfish in the Maldives (Source Photographed by Jan Derk in March 2006 in Fihalhohi, Maldives; released into the public domain)

We find complexity in the behavior of organisms. Swarm behavior of insects and birds, for example, displays *emergent* properties; an example is the way starlings flock together, as shown in the figure on the next page.[2] Such properties demonstrate learning and adaption, as well as precision, all of which we attempt to ascribe to natural processes occurring through evolution over eons of time. Though we can state many fascinating facts about the biosphere that may spark our interests concerning the creation, the more fundamental question regarding faith and science has to do with life's origins.

The previous chapters covered the minute probability that our universe and the world we live on are suitable for us. Now we will take one more step. How can the inanimate become alive? Apart from the question of how space and time began on their own, the story of life's origins is probably the greatest mystery in science. How do lifeless molecules arrange themselves and then be transformed into one of the most complex of apparatuses—the living cell?

Swarm behavior of starlings (Source: John Holmes / Rail Bridge Swarm of Starlings, Creative Commons License)

Biology has made significant progress over the past 100 years. The increase in life expectancy in humans, for example, is a direct indicator of the level of medicine, health education, housing, and nutrition, as well as our understanding of microbiology: all have significantly increased. In 1901, life expectancy in the United States was 46 for men and 48 for women; today, these numbers have increased by 30 years. In only a century, we have added to our knowledge of life science at every level: the cell, interactions of organs in a body, and even complex group behavior in animals.

Despite the biological knowledge available, there remain holes, not the least of which includes the theory of evolution, and arguably, the more fundamental question of the theory of life's origins. The science surrounding the origin of life is, in many ways, more significant than the discussion of evolution.[3] *Abiogenesis*, the science of life's origins, has a major advantage over the science of cosmology, the study of the origins of the universe. Though we will never be able to re-create another universe, we do have the material available here and now to understand and perform experiments on the origins of life. Theoretically, we should be capable of reproducing life from the inanimate in the laboratory, if that is possible. But we shall see just how hard this is.

We caution the reader before advancing. This chapter is a presentation based on the best science available on abiogenesis. Because life science points to the creation through a long process, though that process is still not completely understood, the author is aware that some of the Christian community (e.g., YEC adherents) will not agree with the following presentation. People sharing old earth views should be more comfortable with the material. That being said, it's the author's opinion that a scientific presentation of abiogenesis demonstrates the very point that we wish to show—the extraordinary emergence of life from nonlife and how difficult it is to imagine this occurring only by chance. What are the best arguments that support naturalistic abiogenesis? Or what are the best arguments that refute it? In either case, what do these positions say about a Creator?

The building blocks of life

Living things are complex, and they must be so. For life to do what it needs to do, complexity is essential. Life's complexity starts at the smallest level: the cell. All living creatures are composed of cells, and the simplest creature, a one-cell animal, must activate all the essential functions, including running its metabolism, eating (extracting energy and growing), excreting, replicating, and responding to its environment. For a cell to complete these actions, it must have parts that work together in perfect harmony. According to Professor of biology Michael Denton, "The cell is the most complex and most elegantly designed system man has ever witnessed." He continues,

> To grasp the reality of life as it has been revealed by molecular biology, we must magnify a cell a thousand million times until it is twenty kilometers in diameter and resembles a giant airship large enough to cover a great city like London or New York. What we would then see would

be an object of unparalleled complexity and adaptive design. On the surface of the cell we would see millions of openings, like portholes of a vast space ship, opening and closing to allow a continual stream of materials to flow in and out. If we were to enter one of these openings we would find ourselves in a world of supreme technology and bewildering complexity,...beyond our own creative capacities, a reality which is the very antithesis of chance, which excels in every sense anything produced by the intelligence of man.[4]

A cell is composed of dozens of components, between five million and two trillion molecules, and a total of 10 to 100 trillion atoms, 1000 times more than the number of stars in the local group of galaxies.[5] This small unit of life takes up a space of about 1 micrometer or 0.001 millimeters. Even with a cursory understanding of the cell, one might have the impression that reproducing such a structure is difficult: indeed, it is.

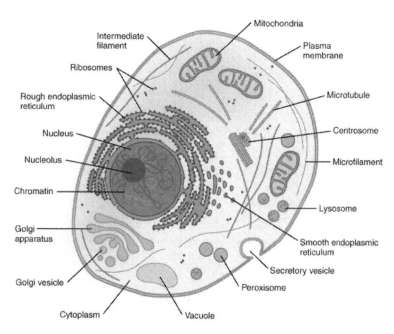

Sketch of an animal cell illustrating the basic parts (Source: withcarbon.com)

The most advanced laboratories are far from reproducing a cell from inorganic molecule building blocks. Given this difficulty, another question comes to mind: Would it be possible to start with something simpler than a cell, something either living or nonliving, and evolve that something into a cell? The answer to this question may indicate just how profound life is on our planet and shed light on the possibility of finding extraterrestrial life. This question is raised not only by the scientific community but also in the faith community. A single animal cell, with its basic parts, is shown in the figure above. The picture only suggests the complexity of the cell but does not do justice to the intricateness of this tiny living machine.[6]

Life in the beginning

The story of the first cell is linked to the early history of the Earth. According to evidence from radiometric dating and other sources, the Earth formed just over 4.54 billion years ago. Material that did not fall into the Sun remained in orbit and slowly coalesced to form our planet. Other planets in the solar system were also created at this time. At the onset, our moon was significantly closer to earth; tidal attraction was considerably stronger, causing disruptions in the mantle, frequent earthquakes, and volcanic activity.[7]

The first half-billion years saw the Earth go through a period of creation reminiscent of Mount Doom from *Lord of the Rings*. The Earth eventually cooled and transformed. Large-scale outgassing and volcanic activity gave rise to the Earth's primitive atmosphere, composed of water vapor, carbon monoxide (CO), carbon dioxide (CO_2), hydrochloric acid (HCl), methane (CH_4), ammonia (NH_3), nitrogen (N_2), and sulfur gas.[8] As water vapor and other gases escaped from the molten rock, the planet cooled, and the oceans formed. The seismic activity on the Earth created great valleys and chasms, and water drained down and filled these holes. It was here in these primeval oceans scientists believe life began.

Life began on earth sometime between 3.5 to 4 billion years ago.[9] Evidently, there must have been a specific time and place where nonlife became life for the first time. What is less clear is whether there were many stops and starts, followed by one last start from which we can trace all living creatures. In any case, a prevalent concept in the field of abiogenesis is the last universal common ancestor (LUCA), considered to be the origin of all living organisms today.[10] Whatever the case, the molecular building blocks available on earth were simply not appropriate for spontaneous creation of a cell, certainly not a cell as complicated as the sketch above; herein lies the problem. If that sketch represents life reduced to its most rudimentary form, then life's building blocks, *nucleotides*[11] and *amino acids*[12] somehow, through a process unbeknownst to us, "came together." If you accept this picture, then we might suggest a metaphor. Imagine a scenario where large piles of brick and mortar, steel and glass, and all the material that makes up a modern urban center, would spontaneously fuse after an energetic pulse (say a lightning strike) and boom, out pops Manhattan. No urban planner would accept this crazy picture to explain the origins of a city; likewise, no scientist accepts the equivalent story of abiogenesis. But the challenge facing microbiologists is to figure out how life could have started, perhaps through many steps and with more rudimentary pre-cell "pieces" and through purely random processes.

Consider the city analogy: a single building did not pop into existence, or a single office space, or even just a desk and chair. The entire Manhattan landscape did not appear all at once; at the start, there was something simpler. From a purely statistical perspective, this simplified picture makes decidedly more sense, but could it work? Can the smaller entity be considered life or just a piece of life? Remember that life must do the following: ingest, evolve, grow, replicate, and sustain itself. If life cannot perform all these functions, then it's not fully alive. That entity must rely on purely nonliving processes to complement what it cannot yet do on its own. As for the urban example, the region of Manhattan has enough sophistication and organization to bring in food, take away garbage, evolve, replace damaged buildings, and thrive; a single desk and chair will have a hard time replicating these functions. Of course, a desk and chair will not evolve unless the other parts of the city are intact. A desk and chair cannot rely on physical processes to become a more complex office space, building, and eventually a city, no matter how much time we allow. Generally, environmental processes surrounding a cell tend to degrade or break down structure. For example, the wind and rain will do their part in reversing the organization of a building, not helping it become more complex (things naturally become more chaotic unless effort or energy is applied—remember Chapter 4). In summary, a subunit of a city (a building or office space) can only function because the city infrastructure is already present; likewise, a subunit of a cell can only function because the rest of the cell supports it. It's difficult to imagine how a fraction of the city or a portion of the cell can function on its own.

RNA and DNA – the brick makers
Before we can use our city analogy further, we must first visit the various building blocks. Life's bricks (cells) are significantly more complicated than building bricks. But there are a few commonalities. At the risk of stretching the analogy too far, we can use similar bricks to build similar buildings. And though cells differ from organism to organism, and from organ to organ within one organism, all cells have certain commonalities. All cells have genetic material, replicating ribonucleic acid (RNA),[13] the more complex deoxyribonucleic acid (DNA),[14] and proteins.[15] There are also common functions among all cells, including those previously mentioned: growth, reproduction, metabolism, and communication or response, as well as *apoptosis* or programmed cell death.

Furthermore, all cells come from other cells. As a side note, though viruses are simpler than cells, they are not living things *per se*, but are complex assemblies of molecules including proteins, nucleic acids, lipids, and carbohydrates. On their own, without living cells, viruses can do nothing.

The *protein* is also a macromolecule and consists of a long chain of amino acids; it can take on different sizes and shapes. Proteins can perform many functions required for life, including transporting molecules from one location to another, responding to stimuli, and catalyzing metabolic reactions, as well as protecting and maintaining cell shape.

All cells have *ribosomes*, complex macromolecular machines that serve as the site of biological protein synthesis. Ribosomes link amino acids together in the order specified by *messenger RNA* or mRNA molecules.[16] The mRNA is essentially a brick producer. The nucleotides form chain links made up of different bases: RNA contains adenine (A), cytosine (C), guanine (G), and uracil (U); DNA contains the same except for thymine (T) instead of uracil (see figure below).[17] Two amino acids link the double helix DNA in what microbiologists call a *base pair*. When linked together, the nucleotides form long strings, either the single-stranded RNA or double-stranded DNA. Together, these large, complex molecules represent the genetic storage units for the cell and are the central key to life.[18]

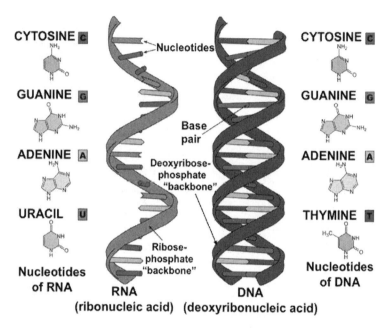

The single-stranded RNA and the double helix structure of DNA— see text for details (Source: Sponk / Wikimedia Commons / CC BY-SA 3.0)

Living organisms cannot perform the above functions without their genetic material, and this is contained in a unit called a *gene*. This distinct sequence contains the genetic material and hereditary information for the cell. A gene ranges in length from about 200 base pairs to nearly 2 million base pairs.[19] All of the essential cellular functions, including the basic instructions for how the entire living organism is assembled and behaves, are stored in the genetic information of the DNA. Cellular functions copy this information to the RNA. The RNA is the equivalent of blueprints in our Manhattan example—every detail needed for the construction, replacement, and functioning of the entire city is found in these blueprints.

The central concept in molecular biology, the "Central Dogma," says that genetic information from the DNA flows to the protein in a two-step process, referred to as *transcription and translation*.[20] First, as the schematic above shows, transcription is the synthesis of an RNA from a DNA. A large catalyst molecule, a protein catalyst or enzyme, opens the double-stranded DNA, usually by about four strands, so that one strand is exposed to the nucleotide. The transcription process creates a template, and an RNA molecule is ready to be formed. Second, translation is performed for protein production from RNA. A host of complex molecules are involved in this production process. Ribosomes link amino acids in a specific order as specified by mRNA. Part of the ribosome molecule "reads" the RNA, and the other parts join the amino acids to form a chain. The ribosomes and other molecules involved make up the translational process. Again, referring to Manhattan, this cellular mechanism is a construct equivalent to carriers, perhaps cars and trucks with work crews, carrying information to a specific location in the cell. When arriving, they perform a needed function, such as fixing a damaged building or office space.

Even in this short description, one hopefully can appreciate the inherent complexity of life at the cell level, which scientists have only begun to understand. We cannot go into more detail here. Still, I encourage the microbiology enthusiast to look into the matter more deeply, starting with the bibliography for this chapter at the end of the book. Many leaps of science have occurred in this field in only the past 20 years. As complex as the translational process is, this critical mechanism is only a small piece in the entire picture of the living cell. The fundamental question is now before us: How could this incredible machine called the cell have started?

The first "steps" of early life

Let us journey to the very beginning when scientists propose that LUCA appeared. Perhaps it was on an afternoon in the shallow oceans or near a thermal vent in the ocean depths. What did it look like? What are the chances that this primitive life could start on its own? What is the probability that natural processes alone were responsible for producing what was needed for its life to begin? To answer these questions, we must first ask how simple first life was. Apart from the creation of the universe, this is the critical place that separates naturalists and creationists. The naturalist believes by fiat that life began only through natural processes, no matter how complex those processes were. Most creationists, with the obvious exception of evolutionary creationists,[21] believe this step too difficult without intervention. We will explore the probability of this occurrence.

A reasonable way to answer this question is to go to the most primitive organism known, one that lives in conditions similar to that found on the early Earth. Bacteria inhabit the volcanic bed in the sea off the shores of northern Iceland. This bacterium is called *Nanoarchaeum equitans*, or roughly, "the riding dwarf." These simple creatures are at the very base of the evolutionary tree and live in the hot sulfur-rich but oxygen-free hydrothermal vents of submarine volcanos. The name dwarf-rider refers to the fact that these ancient forms of bacteria live on a significantly larger *archebacterium*[22] called *Ignicoccuis* (meaning roughly the fire-sphere) in a symbiotic relationship.[23] The temperatures in these waters range from 158 °F to 208 °F or 70 °C to 98 °C, or nearly boiling. The symbiotic pair gains energy by reduction of elemental sulfur to hydrogen sulfide using molecular hydrogen as the electron donor.[24]

The smaller of the pair, the dwarf-rider, measures only 400 nanometers and possesses just under half a million base pairs of nucleic acid. This organism has fewer genes than just about any other organism, so few that it lacks genetic material to control its full energy transport. But because of its symbiotic relationship with its host bacteria the fire-sphere, it can survive.[25] This simple bacterium leaves us with a picture of what life might have looked like 3.5 billion years ago. This bacterium is still far too complex to have randomly popped into existence. So how did it happen?

Over the past century, the science of abiogenesis has been attempting to explain the natural progression and essential steps at the beginning. There remains no single acceptable model to explain the origins of life. The prevailing scientific hypothesis is that life sprung from nonlife in steps. Those steps are characterized by a gradual process of increasing complexity that involved molecular self-

First Life | 129

replication, self-assembly, autocatalysis, and the emergence of cell membranes.[26] We give a rough outline of the steps:

1) The Earth's surface was originally sterile
2) Organic molecules were created by natural processes[27]
3) Primitive cells were formed[28]
4) The biosphere became more complex[29]

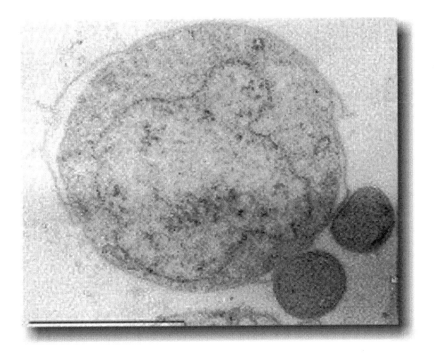

The fire-sphere and two dwarf-riders (Source: Image by Karl O. Stetter and in the public domain)

In the pursuit of validating these steps, scientists have discovered some of the most daunting life-complexity revelations imaginable; the reader is wise to consider them.

In 1952 a graduate student at the University of Chicago, Stanley Miller, persuaded his advisor Harold Urey to perform a relatively simple experiment. With some coaxing, Dr. Urey agreed. Miller and

Urey tested the "primordial soup" hypothesis, an idea that had been presented a few decades earlier.[30] The idea was that the ocean and atmosphere of the primitive Earth were supposedly ideal for life to begin. Miller combined ingredients that he thought were present at the time: steam, methane (CH_4), ammonia (NH_3), and hydrogen (H_2). He exposed the gaseous mixture to an electric discharge that simulated lightning and heat in the early Earth's atmosphere, three and a half billion years ago. After about one week of continuous sparking, Miller discovered a brown broth of amino acids at the bottom of the glass enclosure. This experiment, shown in the sketch below, produced organic chemical compounds using only inorganic molecules.

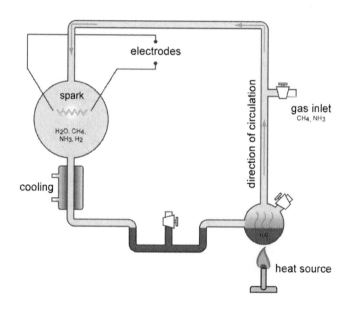

The Miller-Urey experiment setup (Source: Public & Wiki Commons)

Scientists were fascinated by the result: an experiment that reproduced the building blocks of life from inorganic material! But as exciting as this was, over time, the Miller-Urey discovery began to deflate. The first problem with the experiment was that it reproduced conditions that probably did not exist in early Earth history. The current scientific consensus describes the primitive atmosphere not as reducing (oxygen-poor) but either weakly reducing or even neutral.[31] With oxygen present, the atmosphere could not produce amino acids at the rate initially predicted by the Miller-Urey experiment. Scientists realize that oxygen presents a considerable problem in the early atmosphere; so problematic is this issue that alternative sites to the origins of life have been considered, including deep ocean thermal vents[32] and even outer space.[33]

The second challenge of the Miller-Urey experiment is that the step toward creating life was such a tiny one. Though a small step in science is not a disqualifying factor for it being important, the size and complexities of the building blocks should be considered. Amino acids produced in the glass flask of the 1952 experiment contained about 10 to 20 atoms, certainly larger than the two to five atoms of methane, ammonia, and hydrogen that formed them. But the next larger building blocks of living cells are considerably larger. All living matter is composed of nucleic acids, as well as proteins and carbohydrates, as previously mentioned. DNA is composed of nucleic acids and is far, far more complex than an amino acid, containing some 200 billion atoms.[34]

Every cell has DNA, and even the dwarf-rider mentioned above has a fully functional translational system. The analogy of a city again comes to mind. Notwithstanding atmospheric composition, to make amino acids by reproducing "natural" conditions is analogous to tossing a cup, a coaster, and a coffee table in the middle of the floor of an office and hoping that the table lands right side up and the cup and coaster land on the right spot on the table. Maybe all land in the right way and a friendly coffee setting is ready for enjoying, but how rare this would be. Nonetheless, we are still a long way from building Manhattan or even a building, for that matter. In brief, there remain significant barriers to assembling small molecules to produce bigger ones under realistic conditions.

For DNA to replicate, proteins or enzymes must be present, but those same enzymes are produced only at the direction of the DNA; each depends on the other. Cyclic enzyme production is possible in the laboratory, a process in which:

>...the first enzyme binds the two subunits that comprise the second enzyme and joins them to make a new copy of the second enzyme; while the second enzyme similarly binds and joins the two subunits that comprise the first enzyme. In this way, the two enzymes assemble each other—what is termed cross-replication.[35]

The process requires enzymes to begin, and a steady supply of other enzymes to continue. Molecular information is carried on from one cycle to the next, and unless the environment surrounding this enzyme cycle were protected, it would soon end. This brings us to the central issues involving the start of life.[36]

Complex reducibility or irreducible complexity?

To simplify the challenge of reproducing DNA, scientists have speculated that RNA was a precursor molecule to DNA. In other words, life on early earth had self-replicating and catalytic RNA but no DNA or protein. This hypothesis, generally accepted among molecular biologists, is known as the RNA world. Imagine a large complex component in our Manhattan metaphor, maybe One World Trade Center, representing DNA, and a simpler precursor, the Empire State Building, representing a slightly less complex RNA precursor.

As an older building in a modern city, the single-stranded RNA molecule might have performed all the duties in the ribosome, but would eventually be replaced by protein molecules. This evolution would be somewhat like older buildings performing many of the same functions as newer buildings: providing office space, water, electricity, and services.[37] But one must not forget, the RNA world is speculative, and further, relying on RNA does not significantly reduce the complexity. Our coffee table, complete with coffee, is still a very long way from the semblance of the Empire State Building. No matter how many electric shocks zapped the "primordial soup" or how much time elapsed in the depths of the ocean, RNA would simply not just appear. But perhaps a simpler version of RNA existed previous to RNA? And perhaps the complex translation process was replaced by a natural one? These are both significant assumptions, but let's entertain that possibility.

Many molecular biologists believe that natural processes, random polymer synthesis, and natural selection are responsible for the creation of large nucleic acids. The size and shape of these large molecules undergo a process not too dissimilar to the evolution of species, not by breeding but by physical processes including gravity, osmosis, and solvability. If a nonliving membrane were somehow able to form and trap random polymers, then perhaps a large enough polymer could be constructed before the genesis of life. Eventually, there may have appeared a polymer that was large enough to be self-sustaining, occasionally dividing and growing in such a way as to start building its surroundings. At this same time, a simplified (nonliving) energy system would have to exist.[38]

These hypothesized steps were explored in a classic experiment in 1993 by David Bartel and Jack Szostak. Their goal "was to see if a completely random system of molecules could undergo selection in such a way that defined species of molecules emerged with specific properties."[39]

Their experiment did demonstrate this hypothesis.[40] Bartel and Szostak put trillions of RNA molecules in a solution, each one composed of about 300 nucleotides arranged in a random sequence.[41] Depending on their arrangement, some RNA in the experiment could catalyze or react with other RNA better than others. The more reactant RNA selected out other RNA that likewise were more reactant. In 10 series of these experiments, Bartel and Szostak were able to isolate RNA so that it could react seven million times faster than the slower-reaction RNA. Molecular biologist Dave Deamer puts it, "This is the same basic logic that breeders use when they select for a property such as coat color in dogs."[42]

The Bartel and Szostak experiment was lauded as another great leap in science. But as is often the case, when the science of abiogenesis explores one aspect of creation, assumptions must be made about other aspects that cannot be tested.

The natural selection of RNA, identified by Bartel and Szostak, is exciting, but the concept is not new; this process works in many different systems. It's also important to note that not everything needed in the nexus of abiogenesis must be complex. Natural processes do allow for a simple mechanism for self-selection. If rocks and pebbles of different sizes are put in a jar and shaken, the material self-selects: the smaller stones move toward the bottom of the jar, and the larger rocks stay on top. But no matter how long shaking goes on, more complex structures will not appear. By way of analogy,

Michelangelo's statue of David will not appear in a jar, no matter how long you shake it. (I would argue that a simple clay Christmas ornament that I crafted when I was six years old would not appear in the jar.) But the amount of informational content of a Michelangelo statue is orders of magnitudes less than that contained in the RNA molecule, let alone a cell. By shaking a jar, you inhibit the more complex structure.

Similarly, a molecule undergoing too many reactions does not necessarily become more and more complex. The target function of the Bartel and Szostak experiment was self-replication. But this experiment did not explore how complexity is increased; it certainly did not explore how the *correct* complexity is reproduced in cellular functions. We take note that the translational process coexists with multiple other functions, all of which must be coexisting and functioning within a cell.

The first assumption in the Bartel-Szostak experiment was a big one: trillions of potential RNA floating in a calm environment ready to be catalyzed. Scientists believe this was not the case.[43] The second assumption is also a big one. The experimenters selected only one activity or *target function* in the experiment (self-replication), but life processes require many functions that must develop simultaneously. Also, as noted earlier, fast reaction as a target function could potentially be detrimental to the whole process.

Again, at the risk of stretching the analogy too far, we go back to our city example. Many tables with many coffees poured into cups get us no closer to the Empire State Building than before. But the problem is even more profound. The production of a cell membrane is independent of the self-replication process of RNA described above. So unless there was another way to protect the RNA, producing a protective membrane would have to be a target function.

The 300-nucleotide chain of RNA in this experiment required only a few specified nucleotides. What if all 300 nucleotides, populated by four types of amino acids, were specified to perform a function? It would require 4^{300} RNA to be present for the target function to achieve its tasks at any reasonable rate. Any number less than this would mean some activities would be performed, but others not. For our city, this would be as if some utilities were functioning, but others not. Imagine water service available to a city, but no sewage or electricity. Worse yet, imagine sewage service available, but pipes not fully connected or even available. Suppose natural selection (trial and error) produces a minimally functioning life form. While this scenario is en route, detrimental effects would damage any positive steps. Similarly, a city with no sewage or insulation around its electric wires will not kindly wait for missing parts and services that have not yet arrived; the city will either burn up, stink up, or both.

Our Manhattan example has its limits in this analogy. A cell must accomplish *all* the minimal functions of life, including growth, reproduction, metabolism, communication or response, and apoptosis, with a minimal set of elements. A minimal city is not a good analogy here. Imagine a small village or campsite representing the first inhabitants to New Amsterdam on the peninsula of New York. Centuries later, that village would evolve into Manhattan. The problem with this comparison is the difference in irreducible basic building blocks of the two examples. New Amsterdam has people; the cell has amino acids. The lack of services, including electricity, water, and sewage, can be substituted by lanterns, buckets, and a ditch by the savvy pioneer. It's not apparent that a "pioneer" RNA or even protein exists. There remain many challenges if we expect flexible substitution of functions with nonlife processes in early RNA.

Humans are clever "building blocks," significantly smarter than bricks; as a matter of fact, they can even move bricks around and are capable of making adjustments to "evolve" a small pioneer village into a megacity. People can avoid missteps such as uninsulated electric wires or open-ditch sewage through appropriate planning. Natural abiogenesis requires similar functional steps, but "intelligent planning" is not available.

Let's explore the jump from nonlife to life by brute force. If we convert the above number 4^{300} to base 10, it becomes 10^{180}. Recall, there are about 10^{86} elementary particles in the known universe. Even if RNA filled the known universe and all the nucleic acids were allowed to react, we remain hopelessly far from obtaining the goal: an ensemble of target functions to reproduce life's critical characteristics does not have enough time. This fact remains true even if the RNA could react hundreds of times per second since the beginning of time 13.8 billion years ago.

Further, if nucleic acids are not protected, then trial-and-error processes are hampered. Remember, the cell membrane must also be maintained, if not created, by one of the target functions. While random processes are assembling a delicate amino acid, the environment is not optimal. This is especially true if the environment is a hot, deep-sea thermal vent.[44] Even if "smart" automatons inhabited New Amsterdam and attempted to evolve the village to Manhattan, imagine them doing so under constant hurricane conditions.

In summary, we have used a city analogy to explain a cell. We run into challenges, even in the attempt to explain some of the most basic components of a cell. We can reduce Manhattan to the scale of New Amsterdam in 1625. The small settlement can perform the same functions as a modern city but using rudimentary services and devices: lighting by lanterns, water via buckets, and sanitation in outhouses, etc. But the analogy breaks down when we consider the irreducible basic building blocks and planning behind each example. For New Amsterdam, humans are the planners and builders; they are complex and intelligent. There is no natural equivalence of planners and builders concerning RNA, DNA, and the cell as a whole, at least not under our current understanding.[45]

In brief, the Bartel and Szostak experiment observed molecular evolution, but this by no means explains the very challenging probabilistic issues raised in this chapter.[46] It's safe to say that there are many missing pieces of knowledge in the science of abiogenesis.[47]

Where to now?

We could write many pages on life science, but the underlying message is clear—undirected, natural abiogenesis currently cannot be explained. Next, if we gather the summaries from the past three chapters, we can state the following: we currently know only one universe and only one place that harbors life.[48] Only on this Earth has life achieved consciousness, become self-aware, and been able to reason. These are the facts, and scientists do not expect these statements to change anytime soon. So where shall we go from here? Questions surrounding our origins are challenging, but the sincere, curious, and determined scientific mind will make every effort to pursue these questions. So what would a scientist do at this point?

First Life | 137

Here ends our scientific voyage of the physical realm, but not our scientific journey. A true scientist would not be closed-minded to the realm of all possibilities. The *spiritual world*, a world that we cannot see physically (at least directly), should not be out of the reach of questions for the investigator. After all, many of the pursuits in science fall under the category of "that which cannot be seen." The bona fide scientist will keep asking questions about the nature of things, even things one cannot observe directly. As with any other scientific inquiry, the investigator must find other ways to measure that which cannot be measured. A scientist does this by posing questions, examining the evidence, testing the principles, and deciding upon the evidence. As we go to the next chapter and open the Bible, we will use these same tools to explore faith.

Chapter summary and questions

Cells are the simplest form of life and are far too complex to have come about through a spontaneous act in nature. The critical molecules in cells, DNA and RNA, are both unimaginably complex. Some believe that nonlife processes could have naturally allowed for abiogenesis, but there is no substantial evidence to support this. The existence of life is yet another example supporting a "miraculous" intervention from a Creator.

Questions to ponder:
- Can you explain what abiogenesis is in your own words? Though we have not explored evolution in this book, would you say that abiogenesis is a more fundamental topic than evolution in the debate between naturalism and creationism? Why or why not?
- Suppose scientists create a living cell in the laboratory—this may happen one day. The newspapers might read, "The mystery of life's origins solved." What questions should you ask yourself as you want to understand the details of this story?
- What are the best arguments that support naturalistic abiogenesis? What are the best arguments that refute it? In either case, what do these positions say about God (if anything)?

NOTES

[1] Michael S. Rappé, Stephanie A. Connon, Kevin L. Vergin, and Stephen J. Giovannoni, "Cultivation of the ubiquitous SAR11 marine bacterioplankton clade," *Nature*, 418:6898 (2002): 630–633.

[2] *Emergence* occurs when an entity (a school of fish) is observed to have properties its parts do not have on their own (a single fish).

[3] For a good article on the evolution and the origins of life, see Dennis Venema, "At the Frontiers of Evolution: Abiogenesis and Christian Apologetics," BioLogos, https://biologos.org/articles/series/evolution-basics/at-the-frontiers-of-evolution-abiogenesis-and-christian-apologetics, June 19, 2014, retrieved on 25 September 2019.

[4] Michael Denton, *Evolution: A Theory in Crisis* (London: Burnett Books, 1985).

[5] A plant cell could also have been shown without any loss of complexity.

[6] Details of these parts are outside the scope of this book but can be found at the WithCarbon.com site, https://www.withcarbon.com/2014/12/13/animal-cell-parts-and-functions/.

[7] G. Brent Dalrymple, *The Age of the Earth* (Stanford, CA: Stanford University Press, 1991), 492.

[8] J.F. Kasting, Earth's early atmosphere," *Science*, 259 (1993), 920–926.

[9] A. Gutiérrez-Preciado, H. Romero, and M. Peimbert, "An Evolutionary Perspective on Amino Acids," *Nature Education* 3:9 (2010): 2.

[10] D.L. Theobald, "A formal test of the theory of universal common ancestry," *Nature*, 465:7295 (May 2010): 219–22.

[11] A nucleotide is any group of molecules that, when linked together, polymerize to form the building blocks of DNA or RNA. R.J. Young, *Introduction to Polymers* (London: Chapman & Hall, 1987).

[12] Amino acids are organic compounds containing amine (-NH2) and carboxyl (-COOH) functional groups, along with a side chain (an atom or a group of atoms). Cambridge Dictionaries Online (Cambridge University Press), retrieved 9 May 2019.

[13] RNA is a polymeric molecule essential in various biological roles in coding, decoding, regulation, and expression of genes or the sequence of nucleotides that determine the characteristics of the organism.

[14] DNA is a molecule composed of two chains that coil around each other and form a double helix. DNA carries the genetic instructions for cell growth, development, functioning, and reproduction. This is true for all known organisms as well as many viruses.

[15] Proteins are large biomolecules, or macromolecules, consisting of one or more long chains of amino acids.

[16] Messenger RNA (mRNA) is a large family of RNA molecules that convey genetic information from DNA to the ribosome, where they specify the amino acid sequence of the protein products of gene expression.

[17] David L. Nelson and Michael M. Cox, "Nucleotides and Nucleic Acids," in *Lehninger Principles of Biochemistry*, 3rd ed. (New York: Worth, 2000). In addition, there are about 500 naturally occurring amino acids, and each one is composed of only a few elements, including carbon (C), hydrogen (H), oxygen (O), and nitrogen (N).
[18] There are two main differences: the compositions of ribose (sugar) forming the backbones of these molecules and the difference in the base thymine in DNA in place of uracil in RNA.
[19] NIH, National Institute of Health, Genetics Home Reference: Your guide to understanding genetic conditions, https://ghr.nlm.nih.gov/primer/basics/gene, last accessed May 12, 2018
[20] "Message RNA description and the Transcription and Translation process," The National Human Genome Research Institute, https://www.genome.gov/about-genomics, last accessed, May 2020.
[21] To be sure, even evolutionary creationists view God as having a role at the onset of creation. God must have established the proper framework so that the genesis of life "randomly" occured.
[22] Archebacterium: A bacteria-like organism in the kingdom Archaea, kingdom of single-celled microorganism, or an ancient type of bacteria.
[23] Stefan Anitei, "The Smallest Genome: What's the Minimum DNA Amount for Life?" Softpedia News (13 December 2007) https://news.softpedia.com/news/The-Smallest-Genome-What-039-s-The-Minimum-DNA-Amount-for-Life-73763.shtml, accessed June 2018.
[24] H. Huber, M. J. Hohn, R. Rachel, T. Fuchs, V. C. Wimmer, and K. O. Stetter, "A new phylum of Archaea represented by a nanosized hyperthermophilic symbiont," *Nature* 417 (2002): 63–67.
[25] Smaller bacteria have been discovered, possessing only 160,000 base pairs, but their living status is in doubt.
[26] Guenther Witzany, "Crucial steps to life: From chemical reactions to code using agents," *Biosystems* 140 (2016): 49–57.
[27] The naturalistic explanation is that this happened via molecular self-assembly. Also, at about this time (3.5 to 4 billion years ago), the Earth's atmosphere was thought to be reducing or oxygen poor. However, recent research indicates that the atmosphere might have been only weakly reducing or even neutral. H.J. Cleaves, J.H. Chalmers, A. Lazcano, et al., "A Reassessment of Prebiotic Organic Synthesis in Neutral Planetary Atmospheres," *Origins of Life and Evolution of Biospheres* 38:2 (April 2008): 105–115.
[28] Cell membranes are thought to have formed with amphiphilic properties. This property includes both hydrophilic (dissolves in water) and lipophilic (dissolves in fats, oils, and lipids) characteristics. Nucleotides are thought to have polymerized to form RNA and /or DNA and amino acids to have polymerized to form proteins.
[29] Self-assembled structures use energy to accumulate simpler molecules from the environment and then assemble them into reproductions of the original structure. Macromolecular structures become organized into systems within the cell, and complex processes evolve, including replication, transcription, and translation.

[30] The Oparin-Haldane primordial soup concept: simple organic compounds, when exposed to energy, produced organic compounds. See, for example, J.D. Bernal, *The Origin of Life*, reprinted work by A.I. Oparin originally published 1924, Moscow: The Moscow Worker, trans. Ann Synge (London: Weidenfeld & Nicolson, 1967), LCCN 67098482.

[31] Cleaves, et al., "A Reassessment of Prebiotic Organic Synthesis."

[32] Although there are difficulties in this argument as well: the hot temperatures do not create an environment where large, complex molecules can form.

[33] *Panspermia*, or life existing through space and distributed by meteoroids, asteroids, comets, and planetoids. This hypothesis does not solve the origin of life question, but just displaces it elsewhere.

[34] Nucleotides, the building blocks of DNA and fantastically small and less complex, took decades for an experiment with conditions simulating the early Earth to re-create them. See Matthew W. Powner, Beatrice Gerland, and John D. Sutherland, "Synthesis of activated pyrimidine ribonucleotides in prebiotically plausible conditions," *Nature* 460 (May 13, 2009).

[35] Tracy A. Lincoln and Gerald F. Joyce, "Self-Sustained Replication of an RNA Enzyme," *Science* 232:5918 (27 February 2009), 1229–1232.

[36] For an excellent presentation of how the various pieces of a cell come together, the reader is referred to John Oakes, *Is There a God? Questions About Science and the Bible* (Spring, TX: Illumination, 2006), 69–73. Dr. Oakes explains the very unlikely possibility that essential components needed for the simplest cell came together by accident. He also explains the different conditions needed for amino acids and carbohydrates, the former requiring a reducing atmosphere, the latter requiring an oxidizing atmosphere.

[37] Harry F. Noller, "Evolution of protein synthesis from an RNA world," *Cold Spring Harbor Perspectives in Biology* 4:4 (April 1, 2012).

[38] David W. Deamer, "The First Living Systems: a Bioenergetic Perspective," *Microbiology & Molecular Biology Reviews*, 61:2 (June 1997), 239–261.

[39] Deamer, 214.

[40] D.P Bartel and J.W. Szostak, "Isolation of new ribozymes from a large pool of random sequences," *Science* 261:5127 (September 10, 1993), 1411–1418.

[41] 300 nucleotides is considered a reasonable lower limit for an RNA molecule.

[42] Dave Deamer, "Calculating the Odds That Life Could Begin by Chance," April 30, 2009, https://www.science20.com/stars_planets_life/calculating_odds_life_could_begin_chance, retrieved June 2018.

[43] Robert Shapiro, "Astrobiology: Life's beginnings," *Nature*, 476 (August 4, 2011): 30–31.

[44] This argument is adapted from Casey Luskin, "Top Five Problems with Current Origin-of-Life Theories," *Evolution News* (17 Apr. 2017), https://evolutionnews.org/2012/12/top_five_probl/, retrieved August 2018.

[45] This train of logic is related to but not the same as *irreducible complexity*, where a single system is composed of interacting parts in which each part contributes to the functioning of that system. This concept, central to the Intelligent Design view (Chapter 1), proposes that certain biological systems cannot evolve (or originate)

by successive small modifications to preexisting functional systems through natural selection. The example given here focuses on the underlying agents and processes needed to arrive at minimal life.

[46] Multiple other technical issues also require attention: How do complex molecules organize themselves in three dimensions (left-handedness verses right-handedness)? How is bonding determined in three dimensions? How does sequencing work (the arrangement of the components of a complex molecule)? Problems also remain as to how many molecules need to come together and function simultaneously for even the simplest life to exist; this problem evades even a way to attach a probability that this occurred via random processes only.

[47] By no means is this statement an obfuscation to "God of the gaps" (see Chapter 9). But currently, there is simply no easy way to reconcile naturalistic abiogenesis.

[48] Robert W. Graham, "Extraterrestrial Life in the Universe" (NASA Technical Memorandum 102363), Lewis Research Center, Cleveland, OH: NASA (February 1990), archived from the original on 3 September 2014.

Part III – ...and a Step of Faith

After examining the evidence for a Creator through science, we now turn toward the topic of faith. We will focus our attention on faith in the God of the Bible. This may seem like an awkward departure in our study since, at first glance, these two topics look entirely incompatible. On the one hand, we have science: a systematic study of the structure and behavior of the natural world, validated through observation and experiment;[1] on the other hand, there is faith: "the complete trust or confidence in someone or something."[2] What makes these two topics seemingly opposites is something left out of the faith part of this equation. As faith is defined, most of us inadvertently omit its prerequisites; this is especially true in terms of biblical faith. To arrive at the dictionary definition, one must first pass through a phase of evidence and testing. But once tested and tried, faith can mature to produce "complete trust or confidence." We should not minimize this first phase of faith. Those with strong, resilient faith have first passed through this phase; for those who have not, their faith is often weak or superficial.

It's this first phase of faith that we shall explore here. As we do this, we will not suddenly drop our scientific thinking. This first phase of faith, after all, has at its base careful examination with healthy skepticism. This part of the book can be viewed as a continuation in our scientific endeavor—weighing the evidence for God. This step is also essential if one strives to have a relationship with God, as the Bible describes. And as helpful as they are, no amount of physics, astronomy, or biology alone will be sufficient to get someone to this relationship without faith and a journey through the Bible.

---

[1] Oxford dictionary definition of science.
[2] Oxford dictionary definition of faith.

Chapter 7 – The Bible Through the Lens of Science

I don't see why religion and science can't cooperate.
What's wrong with using a computer to count our blessings?
– Robert Orben

If the object of our faith is the God of the Bible, then logically, the most fundamental tool to get there is the Bible. Therefore, we first need to examine the Bible to determine if we can use it as a tool for our faith and, if we can, how it should be applied. As with other scientific endeavors, the Bible suggests that an inquiring mind is essential to seeking and finding a relationship with God.[1]

Our study is on the Christian Bible, a set of 66 books that Christendom regards as divinely inspired and thus constituting Scripture. The authorship and duration to assemble the Bible is in itself a noteworthy piece of evidence. The Bible was written over 1600 years by 40 different people, in 13 countries and across three continents: Asia, Africa, and Europe. There was no committee over the centuries designated to manage its content and guide the authors so that they stuck to a central theme. But the Bible does have one central theme: the fall of man's relationship from God and the recovery of that relationship through Jesus Christ. With such a long list of authors and such an extensive interval of time, it's understandable that the Bible takes on different forms of literary genres, eight to be precise. The different genres offer a rich way for one to understand God; however, the reader must also take the time to understand the context of each of them.[2]

Another surprising fact about the Bible is the breadth of authorship. The writers have a variety of backgrounds, including shepherds, fishermen, soldiers, princes, kings, and doctors. Some were poor, others extremely rich; some were uneducated, others were educated in the best training quarters that society could offer at that time.[3] Despite their background and education, the Bible writers nonetheless succeeded in completing the work without distractions from the central message. Over the sixteen centuries that the Bible writers toiled at their task, kingdoms and civilizations rose and fell, cultures changed,[4] and languages appeared and disappeared.[5] Imagine reading a book with a common theme that was started in the fifth century CE (at the fall of the Roman Empire) and finished this year! The compilation of the Bible was no small achievement.[6]

The Bible covers several subthemes. The first book, Genesis, is about origins: origins of the earth, origins of humanity, origins of sin, origins of languages, and origins of government. The books that follow Genesis cover, for example, the rise and progression of the nation of Israel, its captivity in Babylon, and its return to the land of Canaan. Subthemes in the New Testament (NT) include the incarnation of Christ, the start of the church, and the evangelization of the Roman Empire.[7] The Bible is also well connected—many books prophesy events that occur centuries later. The most significant prophecy in the Bible is the death, burial, and resurrection of Jesus.

Interestingly, the Bible writers foreshadowed this central event in the Bible as early as Genesis 3:15, written 14 centuries earlier. The prophecy of Jesus is then interwoven throughout the Old Testament (OT) by various authors under various circumstances.[8] These prophecies came to a climax in the story of Jesus Christ. His life is presented in the first four books of the New Testament, also known as the Gospels or the good news.[9]

God gets our attention through his words, but how?

Christians sometimes say that God speaks to people in two ways: through his works and his through words (see Chapter 1).[10] The first part of this book covered the creation, God's works; now, we will examine his words. There are new challenges we face here that were not present in the earlier chapters. Previously we dealt with numbers; now we deal with words, and people must interpret words. This additional step may seem like a hindrance, but interpretation is not a bad thing, and it's through that process that we understand what we are reading.[11] We also need to make sure that we use the Bible for its intended spiritual purposes. This detail may seem completely obvious for someone not in the faith community, but the Bible can be misused or miscategorized.[12]

We want to mitigate any misunderstanding of the Bible in terms of its potential scientific insight (or lack thereof), and we present a nonconcordist and literary view of the Genesis account. Alternative views (Chapter 1, and in particular note 35) will provide a different perspective. However, this presentation may be more helpful for those who come from a nonreligious background and are interested in understanding how the Genesis creation account corresponds to the physical creation from a scientific perspective. This presentation does not mean that one is bound to this perspective—views change, and we change (remember those mountain paths of Chapter 1). But for now, we take that first step into the Bible.

The Bible was not written in a vacuum

Bible writers did not produce their works in a vacuum or live in a place void of culture and context. On the contrary, to interpret the meaning of the text correctly, culture and context must be considered. *Theology* is the study of the nature of God and religious belief. In many ways, this discipline is like doing science,[13] in that both require systematic examination of the subject matter, and both call for tests to evaluate the subject. The reader must give thought and put in the effort to understand not just what the writer wrote, but how, when, why, and where the writer wrote it.[14]

We must not forget that the Bible was written by people and for people of many cultures and over thousands of years. As previously mentioned, some 40 people wrote the Bible with the purpose of revealing God's truth to humankind. The investigator should understand that they can read some Bible literature at face value (historical accounts) or even literally. Other text must not be taken literally, for example, biblical poetry. In all cases, regardless of the literary genre, the Bible is clear on one point: all Bible text is inspired by God (e.g., 2 Timothy 3:16–17). We must consider Bible context, especially in terms of our discussion of science and faith. The Bible authors wrote to a contemporary audience, not to twenty-first-century scientists. The student of the Bible should take these words of Longman and Walton seriously:

> Even though the Bible is written for us, it is not written to us. The revelation it provides can equip us to know God, his plan, and his purposes, and therefore to participate with him in the world we face today. But it was not written with our world in mind. In its context, it is not communicated in our language; it is not addressed to our culture; it does not anticipate the questions about the world and its operations that stem from our modern situation and issues.[15]

The message here may seem evident, but for those immersed in Bible culture, we can easily carry Bible authority beyond the spiritual message. We should resist this temptation. Not only was the OT written for the ancient Hebrew culture (the Israelites), but that culture also did not exist in isolation. The Hebrews were very aware of the stories and myths of surrounding pagan societies. God did not call people to forget those stories, though he demanded that his people separate themselves from pagan religious practices (e.g., Leviticus 18).[16] Just as we are familiar with popular stories of our day,[17] so were the Hebrews familiar with pagan myths of creation accounts from neighboring nations. It was in this context that the Genesis writer penned the creation story.[18]

The Genesis account was revolutionary, theologically speaking. One God (monotheism) was introduced, contrasted with the pagan gods and goddesses fighting among each other.[19] The God described in Genesis was responsible for the entire creation, but more importantly, he wanted a personal relationship with humankind. Genesis was juxtaposed with the pagan concepts of gods. Contemporary civilizations created gods in their own creation stories, and humankind was here to serve these gods. The first book of the Bible presented an entirely new view. Yet the Genesis account was framed so that the Hebrews could understand the salient message: the writer did not overhaul the inaccurate science nor the backdrop of the ancient worldview for them.[20]

The state of the ancient Hebrew world

The ancient Hebrews were familiar with creation stories from the Egyptian, Canaanite, Hittite, Assyrian, and Babylonian civilizations, to name a few. Creation stories from these cultures were rich, fanciful accounts. They involved gods and goddesses embodied with human characteristics and weaknesses. Quite often, these gods and goddesses were dependent on humans to serve them. In the effort to contrast the God of Genesis with pagan gods, the author focused on his qualities. Perhaps to reduce the distractions from this comparison, the backdrop of the prevailing worldview was used.

For the ancients, the Earth was flat; it had a boundary (e.g., "the ends of the Earth"). Ancient people thought that water would run off the edges of the Earth had it been spherical. Mountain ranges solved this problem at the Earth's edges. Isaiah gave a brief comment on the shape of the Earth (note a two-dimensional circle, not a three-dimensional sphere) in the following passage: "It is he who sits above the circle of the earth, and its inhabitants are like grasshoppers" (Isaiah 40:22a).[21]

On top of the Earth were the heavens, and they were upheld by pillars (Job 9:6)[22] and a solid dome or *firmament*.[23] The firmament separated the waters of the Earth (below) and the heavens (above). The firmament had to be strong and rigid to support the full weight of the water above. The celestial bodies, including the Sun, the moon, and the stars, were all attached to the firmament. These bodies were "set" in the firmament, not floating above, under, or near it (Genesis 1:17). This understanding was in line with two-dimensional objects, circles for the Sun and moon, and one-dimensional objects, stars. This was consistent with a flat-Earth view. The Earth itself was the center of the universe.[24] This view is consistent with observation—the Sun moved across the sky (Ecclesiastes 1:5; Psalm 19:6; Joshua 10:13). Below was the immovable Earth (1 Chronicles 16:30; Psalm 93:1, 104:5), upheld by pillars. Job, for example, describes earthquakes with imagery taken from this worldview: "He shakes the earth from its place and makes its pillars tremble" (Job 9:6 NIV).

Water played an important role in this ancient worldview. There were two bodies of water, one above and one below the firmament (Genesis 1:6–9). People understood the oceans above the firmament as the place of disorder. The Hebrew words for "void" and "nothingness" have parallel uses in many OT passages and "generally refer to watery chaos." [25] Large expanses of water (oceans) surround the Hebrew world, not empty space. Many ancient civilizations envisioned the concept of the Earth surrounded by oceans. The ancient Hindus, for example, believed that "the earth was spread upon the cosmic waters" and that these primeval oceans "surrounded heaven and earth, separating the dwelling-place of men and gods."[26] In the Babylonian creation epic *Enuma Elish*, the sky is made from the body of Tiamat, the goddess of watery chaos.[27] The Babylonian account describes the god Marduk and his interaction with the waters:

> [Marduk splits] her like a shellfish into two parts: half of her he set up and ceiled it as sky, pulled down the bar and posted guards. He bade them to allow not her waters to escape.[28]

The region below the earth had its purposes. The waters below were a place, not only of chaos but also of fear.[29] The place of the dead, *Sheol*[30] or the underworld, was beneath the Earth. The entire concept of this universe was, in a sense, suspended in nothing, not because there was a notion of balance between centrifugal force and the sun-earth gravitational bond, but (most likely) because there was no understanding of what was outside this picture. The figure by George Robinson, a noted biblical scholar of the early twentieth century, connects the ancient view of Earth with passages in the Bible.[31]

The ancient worldview described here did not dissolve after the Genesis creation account first appeared—it persisted for centuries. For example, other places in the Bible indicate a two-dimensional circle (Isaiah 40:22) rather than a three-dimensional sphere. Also, the idea of the firmament, for example, is used throughout the OT (Genesis 1:6–8, 14–20; Job 22:14, 37:18; Ezekiel 1:22–26, 10:1). In the book of Amos, the firmament or vault is mentioned:

> …who builds his upper chambers in the heavens,
>   and founds his vault upon the earth;
> who calls for the waters of the sea,
>   and pours them out upon the surface of the earth—
> the Lord is his name. (Amos 9:6)

The firmament was a way for people to understand the sky. Biblical scholar and theologian Pete Enns remarks:

There is another approach that attempts to reconcile Genesis and modern science. This approach distinguishes between what ancient authors described and what they actually thought. This is sometimes referred to as the 'phenomenological' view. It acknowledges that the *raqia* in Genesis 1 is solid, but the Israelites were only describing what they saw without necessarily believing that what they perceived was in fact real.[32]

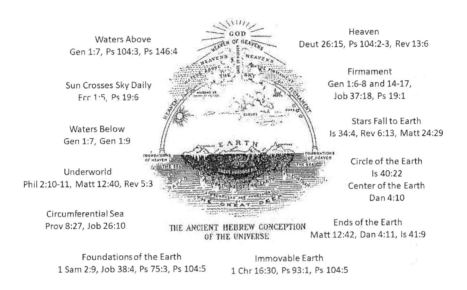

The ancient worldview: The 3-Tier Universe with links to the Bible (Source: Adapted from George L. Robinson, 1913)

This worldview of ancient civilizations is in stark contrast to our modern understanding of the universe, and nowhere in the Bible was an attempt made to correct these misunderstandings.

But should we be surprised by the Hebrews' incorrect view of the heavens? In this author's opinion, not at all. All ancient civilizations had erroneous concepts of the Earth and the surrounding space.[33] The Bible writers used *phenomenal language*: they described things as they appeared. Their cosmology was, therefore, a prescientific attempt to understand the world around them.[34] We use phenomenological language today to describe everyday celestial events. We use the terms "sunrise" and "sunset" as we refer to the apparent position of the Sun with respect to our position on earth. Yet in its six-billion-year existence, the Sun has never risen or set once.[35]

The poetry genre, which covers 27 % of all biblical text, is also an important consideration. We must not confuse descriptive language with scientific notions when the context is clearly poetic. In the book of Jeremiah, for example, a description of God's power is written in poetic language:

> When he utters his voice, there is a tumult of waters in the heavens,
>   and he makes the mist rise from the ends of the earth.
> He makes lightnings for the rain,
>   and he brings out the wind from his storehouses.
> (Jeremiah 10:13)

No one should assume that this Bible passage is teaching about a mysterious building or room in the sky where the weather is stored. Likewise, no one would be concerned about falling felines and canines when someone says, "It's raining cats and dogs outside."

The Genesis creation account

The Bible describes the Genesis creation account in less than 1000 words. The form of the account is *cosmological*,[36] based on other creation accounts from other civilizations minus polytheism and the myths. This creation account does not mean that science was entirely void in the Genesis account; it merely means that it was not the focus. The writer of Genesis attempted to contrast the God of the Bible with the multitudes of foreign gods. Bible historian Conrad Hyers writes:

> Given this as the field of engagement, Genesis 1 is cast in *cosmological* form—though, of course, without the polytheistic content, and in fact over against it. What form could be more relevant to the situation, and the issues of idolatry and syncretism, than this form? Inasmuch as the passage is dealing specifically with origins, it may be said to be cosmogonic.[37]

Moses, the traditional author of the first five books in the Bible, including Genesis, attempted to contrast monotheism and polytheism. He knew that the Hebrews were familiar with the concepts of chaos, water, and the firmament. Had Genesis been framed in terms of modern cosmology, this revolution in science would have blurred the Hebrews' understanding of God.

Given this backdrop of the Genesis cosmogony, it's easier to see an underlying structure in the account. The creation "problem" is presented in three different stages: darkness (Stage 1), watery abyss (Stage 2), and a formless Earth (Stage 3). Corresponding "solutions" are provided for each stage: the creation of light and separation of light from darkness in the form of day and night (Stage 1); creation of the firmament and separation of the water above from the water below (Stage 2); and separation of the Earth from the sea and the creation of vegetation (Stage 3). These three solutions presented at the three stages prepared the way for three different populations. Populating these stages included nonlife objects: the Sun, moon, and stars (Stage 1), birds and fish (Stage 2), and land animals and humans (Stage 3). Each population provided for the next stage. The Sun, for example, provided needed light for life (e.g., birds and fish), which subsequently offered food for the third stage, land animals, and humans. The table summarizes these steps.[38]

Stage	Problem Genesis 1:2	Preparation Days 1–3	Population Days 4–6
1	Darkness (1a)	Creation of light (1a) and separation from darkness – night (1b)	Creation of the Sun (4a), the moon, and stars (4b)
2	Watery abyss	Creation of the firmament (2a) and the separation of waters above from the water below (2b)	Creation of birds (5a) and fish (5b)
3	Formless Earth	Separation of Earth from the Sea (3a) and creation of vegetation (3b)	Creation of land animals (6a) and creation of humans (6b)

The cosmogony of Genesis 1 (Source: Adapted from C. Hyers)

The entire creation event is contained in one week (seven days). This is not an accident. According to Conrad Hyers,

> Seven was a basic unit of time among West Semitic peoples, and goes back to the division of the lunar month into 4 periods of 7 days each.[39]

It can be argued that for religious reasons, not scientific reasons, creation needed to be completed in the natural timeframe, which was based on the natural division of one week in the lunar month ($4 \times 7 = 28$);[40] six days were required in total with an additional seventh for resting.[41]

The end of the matter

Galileo Galilei once quoted his contemporary Cardinal Cesare Baronius as saying, "The Bible teaches us how to go to heaven, not how the heavens go." In other words, Galileo was expressing the sentiment that scientific explanation is not the goal of the Bible and, likewise, not the goal of Genesis. Conrad Hyers again writes precisely to the point about Genesis:

> So the issue is not at all, How is Genesis to be harmonized with modern science, or modern science harmonized with Genesis? That kind of question is beside the point, if by the question one is proposing to try to synchronize the Genesis materials with materials from the various fields of natural science: biology, geology, paleontology, astronomy, etc. That would presuppose that they are *comparable*—that they belong to the same type of literature, level of inquiry, and kind of concern. But they do not. Trying to compare them is not even like comparing oranges and apples. It is more like trying to compare oranges and orangutans.[42]

The Bible was written in a framework so that it was understandable for those reading it at that time. It would have been a hindrance to the Israelites had they been forced to wade through a completely foreign conception of modern science.[43] The very notion of writing the Bible to the twenty-first-century God-seeker has other problems. Our worldview is grounded in twenty-first-century science.

But what about science in 100 years, 1000 years, or 10,000 years? Will future Bible readers not be discussing the same issues we are discussing now, but needing to wade through additional layers of complication? Imagine if the Bible writers wrote the text with our current level of understanding of science. Future Bible readers would first have to translate from the ancient creation story to the twenty-first century, but then from the twenty-first century to their own. And what is so special about our current understanding of science? Why didn't the Bible writers pen the books for the tenth century, the twelfth century, or the Renaissance scientist? The twenty-first-century reader would have been perplexed, and so would the original readers of Genesis.

There are many tools available to a scientist who seeks to understand the physical world, but not all of them are useful. Likewise, there are many tools available to the God-seeker, but not all are helpful. An astronomer would never use a microscope to study the stars; a microbiologist would never use a telescope to study cells. This does not mean that the Bible is unscientific; it only means that its purpose was not for science. The Bible may contain scientific information, somewhat like a science book containing illustrative information from another domain (see Appendix 3). A physics textbook may demonstrate the concept of rotational inertia using a skating couple, male and female, holding hands, but are we to deduce anything about skating, dating, or marriage by this example?[44] The answer: only if we are looking for it. When we insert our own bias into the Scriptures, it's easy to forget that its main thrust is spirituality.

Where do we go from here?

At this juncture, one might have the impression that the Bible is utterly useless for scientific inquiry in terms of reaching an understanding of God. One might even think after reading this chapter that the biblical text is impractical, out of date, or simply primitive. Nothing could be further from the truth. Though the Bible is written with an ancient worldview in mind, it's filled with astounding facts based in history. Unfortunately, these facts are watered down by the relatively recent use of the Bible to answer science questions. Ironically, instead of elevating the Bible as a source of supernatural insight for twenty-first-century science, forcing the Bible to explain science only obscures the discussion.

It's appropriate to summarize this chapter in the words of Longman and Walton, "Even though the Bible is written for us, it is not written to us."[45] The Hebrews were party to theological innovation, not scientific innovation. And though the Bible writers never intended their work as a science tool, it's a brilliant historical work. This is the direction in which we shall go next.

Chapter summary and questions

The Bible is not a science book, nor should it be used to glean scientific information. The account of creation was written in a prescientific cosmogony.

Questions to ponder:
- Explain what this phrase means to you: "The Bible was written for us, not to us."
- If the Bible was written under the assumptions of an ancient (and incorrect) scientific worldview, how can we rely on it as a valid source of information concerning faith in God?
- How would you reconcile the inaccurate ancient worldview of the Earth with the concept of the inerrancy of the Bible?
- Could God have imbedded "hidden" scientific information in the OT or elsewhere in the Bible? Can you think of a challenge (or benefit) to that concept?
- How willing are you to be flexible in your biblical view of creation, especially when assisting a nonreligious (and perhaps scientifically inclined) friend?

NOTES

[1] The Bible is replete with commands and suggestions to study it: Joshua 1:8; Psalm 119:105; Proverbs 3:1–2; Jeremiah 29:13; Romans 15:4; Revelation 1:3.

[2] The different genres include: Historical Narrative, the Law, Wisdom, Psalms, Prophecy, Apocalyptic, Gospel, and Epistle (letters to the churches). We recommend reading Fee, Gordon D. and Douglas Stuart. *How to Read the Bible for All Its Worth* (Grand Rapids, MI: Zondervan, 2003).

[3] The education of authors in the New Testament, for example, varied greatly. The apostle Mark was probably the least educated and Luke or Paul were likely the most knowledgeable. Their messages came together, nonetheless, in a seamless,

rational way. Ehrman, Bart D. *Jesus. Apocalyptic Prophet of the New Millennium* (New York: Oxford University Press, 1999), 45.

[4] Culture is an important consideration in understanding the Bible. Though the Bible message has been written for all time and for all humanity, the writings were directed to a specific ancient culture. This ancient culture was male dominated. Had the Bible been written today, I would expect more books to have been written by women.

[5] A timeline of the world's languages can be found at eLinguistics: http://www.elinguistics.net/Language_Timelines.html

[6] How did the Bible come together? For the interested reader: Neil R. Lightfoot, *How We Got the Bible* (Grand Rapids, MI: Baker Book House, 1988). Also, an excellent audio series on this topic is available by Douglas Jacoby; see the bibliography for Chapter 7.

[7] Peter Pilt, "Compelling Evidence of the Authenticity of the Bible," May 24, 2012, https://peterpilt.org, last accessed November 2018.

[8] In the Protestant canon (authoritative Scriptures), the OT represents the first 39 books in the Bible.

[9] The NT is a collection of 27 books written in Koine Greek and includes the four canonical gospels (Matthew, Mark, Luke, and John), the Acts of the Apostles, the fourteen epistles of Paul, the seven catholic epistles, and the Book of Revelation.

[10] This thought is often stated in reverse: "his words and his works" (the two books of God). But there are other ways God can speak to us, including through Jesus and history (Chapter 8) as well as our conscience, and by council from more spiritual people (e.g., elders in the church). See Copan and Jacoby, *Origins*, 229–230.

[11] This process of interpretation is not necessarily always easy. Apart from crystal-clear scriptures, most of the Bible requires careful homework: pondering, considering, evaluating, weighing evidence, etc. There is an analogy here to a scientific inquiry.

[12] By incorrectly categorizing it, we imply that the Bible has been considered to possess authority not only in the spiritual domain, but in the scientific domain as well—this is a category error. We refer the reader to an enlightening discussion on this topic in Copan and Jacoby, *Origins*, 27.

[13] One might say that science is like attempting to understand God's thoughts and his actions, while faith (the Christian faith in particular) is trying to understand God in terms of a relationship with him.

[14] Applications of the Bible (not the thrust of this book) must also be considered carefully and in context. For example, the Bible urges us to imitate Paul as he imitates Jesus (1 Corinthians 11:1), but that raises the question: Should we do everything Paul did? Should we purchase a ticket to Turkey or Greece, dress up like Paul, and go to the Areopagus and challenge the tourists to follow Jesus? Hopefully, the point is clear.

[15] Tremper Longman III and John H. Walton, *The Lost World of the Flood: Mythology, Theology, and the Deluge Debate* (Downers Grove, IL: IVP Academic, 2017).

[16] Despite warnings from OT prophets, not all practices ceased, and some were not called out until much later. The practice of polygamy, for example, was certainly not mandated by God, though more than 40 important figures in the OT had multiple wives. The original intention was for a man to have one wife (Genesis 2:24), but this practice was not challenged directly until the NT (e.g., 1 Corinthians 7:2–4).

[17] If Genesis were written today, it might be written in contrast to the currently most popular science concept: the Big Bang theory.

[18] We refer the reader to Copan and Jacoby, *Origins*, 11–22, for insights into the cultural backdrop of the time.

[19] This topic is outside our scope; we refer the reader to Evert Fox, *The Five Books of Moses, The Schocken Bible*, Volume 1 (New York: Random House, 2000). Also, a thorough discussion of ancient gods and goddesses can be found in Copan and Jacoby, *Origins*, 31–43.

[20] The salient difference between the Genesis account and the pagan accounts is not the science that's present; it's the mythology that isn't. More of this topic can be discovered in Gordon J. Wenham, *Genesis 1–15*, Word Biblical Commentary, Volume 1 (Waco, TX: Word Books 1987), xlviii. See also Copan and Jacoby, *Origins*, 45–94.

[21] Gier points out that the idea of a spherical earth did not enter Jewish thought until the Middle Ages. N. F. Gier, "The Three-Story Universe," in *God, Reason, and the Evangelicals* (Lanham, MD: University Press of America, 1987), Chapter 13.

[22] It has been suggested that, with a bit of hindsight and perhaps a bit of bias, contemporary Bible readers could assert that foundations or pillars of the Earth (Job 6:9, 38:4–6; 1 Samuel 2:8; Psalm 75:3) must be metaphorical, while the single verse in Job (Job 26:7) is literal. Yet there is no reason to force this interpretation on the Scriptures.

[23] The term "firmament" has been translated "expanse," but this is not an accurate translation and is inconsistent with the original concept of the Hebrews. Pete Enns puts it this way: "Ancient Israelites 'saw' this barrier when they looked up. There were no telescopes, space exploration, or means of testing the atmosphere. They relied on what their senses told them. Even today, when looking up at a clear sky in open country, it seems to 'begin' at the horizons and reaches up far above. Ancient Israelites and others in that part of the world assumed the world was flat, and so it looked like the Earth is covered by a dome, and the 'blue sky' as a result of the 'waters above' being held back by the *raqia*. The translation 'firmament' (i.e., firm) conveys the idea of a solid structure." Pete Enns, *The Firmament of Genesis 1 Is Solid but That's Not the Point*, BioLogos (January 14, 2010) https://biologos.org/articles/the-firmament-of-genesis-1-is-solid-but-thats-not-the-point.

[24] This is the geocentric model as opposed to the heliocentric view originated by Copernicus in the fifteenth Century (see Chapter 2).

[25] Gier points out that these concepts of chaos are also seen in Genesis 1:1, Jeremiah 4:23, and Isaiah 40:17, 23. As an extension, the writer of Job was

probably not thinking of the Earth orbiting the Sun. N.F. Gier, "C. The Waters Above and Below," in "The Three-Story Universe."

[26] The *Rig-veda* (trans. O'Flaherty), 32, 29.

[27] Gier, "The Waters Above and Below."

[28] Ibid.

[29] In Celtic culture, for example, there was fear that the seas would come rushing in from all directions. Charles Squire, *Celtic Myth and Legend* (Mineola, NY: Dover, 2003), 174.

[30] In the Hebrew Bible, Sheol was a place of darkness to which all the dead go, both the righteous and the unrighteous. Sheol, translated in Greek as Hades, was not considered the same as hell.

[31] This illustration is from George L. Robinson, *Leader of Israel* (New York: NY: Associated Press, 1913), p. 2.

[32] See Enns, *The Firmament of Genesis 1*. Also, those interested in more details can begin by reading Paul H. Seely, "The Firmament and the Water Above" (two-part article in the *Westminster Theological Journal* 53 (1991): 227–40 and 54 (1992): 31–46; also, John H. Walton, *Genesis*, NIVAC (Grand Rapids, MI: Zondervan, 2001), 110–13.

[33] This is not to say that our view of the universe is complete.

[34] Even the term "cosmos" had no equivalent in the Hebrew language. "*Kosmo*" was first used by Pythagoras, the first to conceive of the universe as a rational, unified whole. Such a notion is crucial to the scientific idea that things operate according to lawlike regularity. For the Hebrews, the universe was not a cosmos, but a loose aggregate held together and directed by God's will. *The Interpreter's Dictionary of the Bible*, Vol. 1 (Nashville, TN: Abingdon, 1981), 702.

[35] Technically speaking, from an observer's perspective on earth, the position of the Sun creates the appearance that our star disappears behind our planet as we rotate. Thus, it only appears to us that the Sun is rising and setting. This description is not scientific but phenomenological. It's also a completely valid way to communicate.

[36] That is, relating to the origin and development of the universe.

[37] Hyers, Conrad. "The narrative form of Genesis 1," *Journal of the American Scientific Affiliation* 36:4 (1984): 213a. Accessible at https://faculty.gordon.edu/hu/bi/ted_hildebrandt/OTeSources/01-Genesis/Text/Articles-Books/Hyers_Gen1_JASA.htm, 209a.

[38] Hyers, "The narrative form of Genesis 1," 213a.

[39] Ibid.

[40] Hyers puts it, "It would surely have seemed inappropriate and jarring to have depicted the divine creative effort in a schema of, say, 5 days or 11 days." Hyers, "The narrative form of Genesis 1," 213a.

[41] Understanding numerology for the Hebrews is critical in making sense of the text, not only with the topic of the creation but in the subsequent text concerning genealogy. We refer the reader to a very interesting analysis of this topic by Carol A. Hill, "Making Sense of the Numbers in Genesis," in *Perspectives on Science and Christian Faith* (the *Journal of the American Scientific Affiliation*) 55:4 (December 2003), 248.

[42] Hyers, "The narrative form of Genesis 1," 210a.

[43] According to Hyers, "A basic mistake through much of the history of interpreting Genesis 1 is the failure to identify the type of literature and linguistic usage it represents. This has often led, in turn, to various attempts at bringing Genesis into harmony with the latest scientific theory or the latest scientific theory into harmony with Genesis. Such efforts might be valuable, and indeed essential, if it could first be demonstrated (rather than assumed) that the Genesis materials belonged to the same class of literature and linguistic usage as modern scientific discourse," Hyers, "The narrative form of Genesis 1," 213a.

[44] More on this idea in Appendix 3.

[45] Longman and Walton, *The Lost World of the Flood*. Also, Copan and Jacoby put it this way: "To connect with the people of their day, the writers of the Bible had to use words, phrases, and concepts that were familiar to their audience. They may have been writing for posterity, but surely they weren't focused on twenty-first-century issues. Their concern was *their* world, *their* era." Copan and Jacoby, *Origins*, 15.

Chapter 8 – The Bible Through the Lens of History

The consequences of our actions are so complicated, so diverse, that predicting the future is a very difficult business indeed. – J. K. Rowling

Regarding the Bible, there are three questions a healthy skeptic should ask: 1) Did these events happen? 2) Are there historical records? 3) Are the prophecies of the Bible accurate and valid? All three of these questions must be answered in the affirmative if we are to use this book to build our faith.

Of course, you don't have to look far from geographic descriptions in the Bible to know that it connects with reality. Cities, hills, mountains, lakes, and seas are all recorded in the historical accounts of both the Old and New Testaments. More surprising are the recent archaeological discoveries that corroborate Bible accounts. These discoveries have lifted the doubt that overshadowed some Bible stories for decades.

The historicity of the Old Testament

Discoveries are the key to testing ideas and theories in any scientific endeavor. In this view, we will explore the historicity of Bible, or its "acceptability as a history."[1] This investigation includes exploring the many archaeological findings that corroborate the Old and New Testaments. Sometimes, these discoveries follow many years of skepticism. J. Warner Wallace writes about the relevance of archaeological findings and the Old Testament:[2]

> I do not expect a surveillance video confirming every statement made by a witness, but I do expect small

"touchpoint" corroborations.... For example, over 50 people mentioned in the Bible have now been confirmed through archaeological discoveries. While archaeology is notoriously partial and incomplete, it does offer us "touchpoint" verification of many biblical claims, including the claims of the Old Testament.

One of the most important touchpoint discoveries came only a few decades ago. In 1993, the archaeologist Avraham Biran was directing an excavation at Tel Dan, a mound of earth that covered centuries of habitation in the extreme north of modern Israel. The team uncovered a ninth-century BCE stone slab or *stela*. That slab contained writings that gave the first historical evidence of King David, the legendary king of Israel in the tenth century BCE and writer of a considerable portion of the book of Psalms. Ancient Near East (ANE) historians had been very skeptical of a united monarchy under the leadership of Saul, then David and his son Solomon, a period that lasted 120 years. Furthermore, until this recent discovery, there had been no evidence outside the Bible that David existed. That the stone contained the written phrase, ביתדוד, consisting of the Hebrew words "house" and "David," translated "House (Dynasty) of David," is most significant.

An Aramean king erected the stela in the mid-eighth century BCE. Though the stela does not mention his name, it's believed to be Hazael of Damascus, who defeated both Jehoram of Israel and Ahaziah of Judah. These bloody encounters between the divided kingdoms of Israel and the kingdom of Aram to the north are all recorded in the book of 2 Kings, chapters 10, 13, and 19. In the fragmentary description, the Aramean king boasts that he defeated the "king of the House of David," and "under the divine guidance of the god Hadad, vanquished several thousand Israelite and Judahite horsemen and charioteers before personally dispatching both of his royal opponents."[3] The etching on the Tel Dan stone slab has convinced most scholars on the subject and ANE archaeologists that this represented the first concrete proof of the historical King David. Admittedly, the inscription was more about boasting of a victory over Israel during its declining years following the 120 years of unified reign, but it nonetheless provided that touchpoint mentioned by Wallace.

The fragmentary Tel Dan stela, containing the Tel Dan inscription or "House of David" inscription (Source: The Israel Museum, Jerusalem/Israel Antiquities Authority; M. Suchowolski)

Not long ago, the walls around Jerusalem were part of an archaeological discovery. Nehemiah rebuilt the walls in 445 BCE, and the book that bears his name mentioned that that rebuilding took 52 days (Nehemiah 6:15). Historians had a healthy skepticism and doubted the existence of this wall, since it was never previously discovered. It was not until 2007 that an excavation uncovered remnants of the wall in Jerusalem. Archaeologists are now in general agreement that this was indeed Nehemiah's wall and that the adjacent palace found in the same dig was King David's palace.

Connections with nations outside Israel can also corroborate the Old Testament. The height of the Assyrian Empire, also known as the Neo-Assyrian Empire (911–627 BCE), was a period of significant expansion. The capital was Aššur (911 BCE) in modern northern Iraq, but was eventually moved to Nimrod (879 BCE), then Dur-Sharrukin (706 BCE), Nineveh (704 BCE), and finally Harren (612 BCE). At its peak, the empire extended from northern Iraq and central Turkey in the north, toward modern Kuwait at the banks of the Persian Gulf in the east, and Egypt to the south and west. The Levant region, including the divided kingdom of Israel (Samaria) and Judah, was contained within the Assyrian dominion and, at one time, was overtaken, as in the case of Samaria. Biblical accounts of the divided kingdom of Samaria and Judah are numerous, and over the past century and a half, artifacts from the region of northern Iraq have corroborated the biblical accounts. But until discoveries of the nineteenth century, much of the narrative surrounding the late Iron Age in the Ancient Near East (1200–500 BCE) remained unverified.

During the Victorian expansion in the mid-1800s, the youthful Austen Henry Layard, an English traveler and archaeologist, voyaged to the region of Mosul (in modern-day Iraq). Layard had the opportunity to explore the ruins of Assyria, partly excavated by a more senior and well-known French archaeologist, Paul-Émile Botta. While working with Botta, Layard uncovered several large stone reliefs in the ancient ruins of the Assyrian palace in Nineveh. These massive stone panel sculptures were carved between 700–681 BCE as a decoration of the Southwest Palace of Sennacherib, who was the king of Assyria from 705 BCE to 681 BCE.[4]

The 28-year-old Layard studied the reliefs and learned about the narrative. He later wrote in his journal,

> Here, therefore, was the actual picture of the taking of Lachish, the city as we know from the Bible, besieged by Sennacherib, when he sent his generals to demand tribute of Hezekiah, and which he had captured before their return; evidence of the most remarkable character to confirm the interpretation of the inscriptions, and to identify the king who caused them to be engraved with the Sennacherib of Scripture. This highly interesting series of bas-reliefs contained, moreover, an undoubted representation of a king, a city, and a people, with whose names we are acquainted, and of an event described in Holy Writ.[5]

Another famous explorer, Sir Henry Rawlinson (1810–1895), questioned the cuneiform characters representing the scene that Layard identified as the city of Lachish.[6] Over time, however, further archaeological scrutiny confirmed Layard's original identification.

Today, if you are in London and go to the British Museum, you will see the Lachish reliefs, large stone depictions of Assyrian victories over the kingdom of Judah during the siege of Lachish in 701 BCE. The figure shows a reference to Sennacherib, the king of Assyria (705 BCE to 681 BCE) in the top right of the relief:

> Sennacherib, the mighty king, king of the country of Assyria, sitting on the throne of judgment, before (or at the entrance of) the city of Lachish (Lakhisha). I give permission for its slaughter.

The Lachish inscription (Source: British Museum, Creative Commons Attribution)

Sennacherib is also cited numerous times in the Bible, and three places in the Bible specifically connect to this relief. First, in the book of Isaiah, it is written:

> In the fourteenth year of King Hezekiah, King Sennacherib of Assyria came up against all the fortified cities of Judah and captured them. The king of Assyria sent

the Rabshakeh from Lachish to King Hezekiah at Jerusalem, with a great army. (Isaiah 36:1–2a)

Second, in the Book of Chronicles, it is written:

> After this, while King Sennacherib of Assyria was at Lachish with all his forces, he sent his servants to Jerusalem to King Hezekiah of Judah and to all the people of Judah that were in Jerusalem, saying, "Thus says King Sennacherib of Assyria: On what are you relying, that you undergo the siege of Jerusalem?" (2 Chronicles 32:9–10)

Third, in the Book of Kings, in the context where Sennacherib threatens Jerusalem, it is written:

> The king of Assyria sent the Tartan, the Rabsaris, and the Rabshakeh with a great army from Lachish to King Hezekiah at Jerusalem. They went up and came to Jerusalem. When they arrived, they came and stood by the conduit of the upper pool, which is on the highway to the Fuller's Field. (2 Kings 18:17)

Assyrian records mention incidents between the empire, Israel, and Judah, including Shalmaneser III (second king of the Neo-Assyrian Empire). The archaeologist Bleibtreu writes,

> An inscription of Shalmaneser III records a clash between his army and a coalition of enemies that included Ahab, king of Israel (c. 859–853 BCE). Indeed, Ahab, according to Shalmaneser, mustered more chariots (2000) than any of the other allies arrayed against the Assyrian ruler at the battle of Qarqar on the Orontes (853 BCE). For a time, at least, the Assyrian advance was checked.[7]

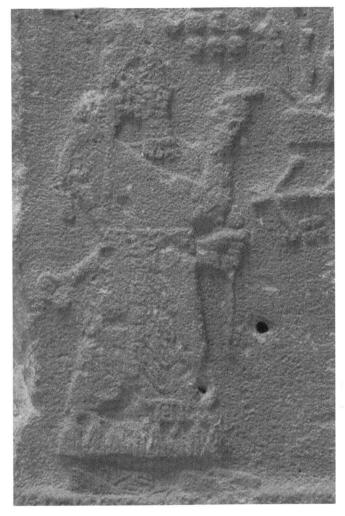

Tell al Rimah Stela (797 BCE): Inscription by Assyrian king Adad-Nirari III, in which he describes his successes in the west and the tribute from Jehoash the Samarian (Source: Istanbul Archaeological Museum, photo by Marco Prins, CCO 1.0 Universal)

A stela from Tell al Rimah (806 BCE) depicts the Assyrian king Adad-nirari III. This sculpture connects us with the OT and informs us that Jehoahaz, king of Israel (814–798 BCE), paid tribute to the

166 | A Step of Faith

Assyrian king: "He [Adad-nirari III of Assyria] received the tribute of Ia'asu the Samarian [Jehoahaz, king of Israel], of the Tyrian (ruler) and the Sidonian (ruler)."[8]

The Assyrian conquest is just one of the many connections between the Old Testament and archaeological discovery. Many other archaeological corroborations span the history of the OT.[9]

The historicity of the New Testament

The New Testament must also be validated if we are to trust its contents and therefore, the central message of the Christian doctrine. The study of the historicity of the New Testament explores this issue. For centuries, both Christian and non-Christian scholars have weighed in with their opinions on the NT. Many of them have focused on the historicity of Jesus. In reviewing this topic, the author has noted that in the specific study of Jesus, there is a considerable tendency for scholars to bring in their predetermined views. But as with any science, we must resist predetermined views and study why the NT, the Gospels, and Jesus can or cannot be trusted. This study is extensive, and though we can only do an overview in this book, I encourage the reader to refer to the extended bibliography for Chapter 7 for a more in-depth study of the topic. Before we look at the scholars' views, we begin by mentioning a universally accepted event. Most scholars agree that Jesus was a historical person, but beyond this statement, scholars tend to diverge in their view of the NT and their view of Jesus. This is where textual criteria are so necessary. Primary historicity standards include validation of authorship[10] and validation of textual consistency.[11] But there are several other tests as well, including validating both internal and external consistency of the text;[12] clarifying intention and genre;[13] establishing the sources from oral traditions;[14] and textual criticism.[15]

Many scholars of antiquity support the idea that the four synoptic Gospels are authentic.[16] Some scholars take issue with the Gospels' authenticity, but generally, they focus their criticism on specific events. For example, they may draw attention to uncertainty in details of the nativity or particulars about the crucifixion.[17]

But there are many arguments on the side of validating the text. The short gap of time between the actual event and the earliest recorded text adds to the authenticity of the Gospels. The earliest known fragment of the New Testament in our possession is dated to only about 40 years after the Apostle John wrote the last letter to the church. Scholars have estimated that John wrote the gospel account (the book of John) and his three letters to the church around 98 CE; the John Ryland fragment dates to approximately 125–130 CE.[18] The first complete copy of a New Testament book appeared around 200 CE, and the oldest NT copy is the Codex Sinaiticus, appearing in the fourth century.[19]

Though these gaps may seem to represent an expansive period, they are relatively short when compared to all other ancient documents in relation to their original work. Plato's dialogue, for example, dates to 1000 years after Plato wrote the original text, yet no credible scholar disputes the veracity of this copy in transmitting Plato's words to us today. Only a few other ancient documents come remotely close to the short time gap associated with the New Testament. The historical account by Pliny the Younger is one example, with the original writing dated from 61–113 CE and the oldest copies dating from 850 CE, spanning a time of about 790 years.[20] Only seven copies of Plato's and Pliny's works exist.

On the other hand, there are 24,000 manuscript copies of portions of the New Testament available to us today. Again, in contrast, the ancient epic poem by the Greek writer Homer, the Iliad, was written between 800–750 BCE; the oldest copy dates to 400 BCE and some 1800 manuscripts survived.[21] All other ancient documents have either fewer surviving manuscripts than the New Testament or longer time gaps in comparison to the NT.[22] Yet despite their smaller numbers, the reliability of the other ancient documents is not seriously questioned.

In terms of validating the text, scholars like to see many copies and a short time gap between event and copy. The more copies that are available, the more scholars can cross-check the work. Also, text generally has fewer errors when there is a shorter gap between the event and the copy. The NT satisfies both of these qualities, and moreover, someone could virtually reproduce the entire New Testament from citations contained in the works of the early church leaders. A fact of note: there are some 32,000 citations in the *Writings of the Fathers* prior to the Council of Nicaea in 325 CE.[23]

Another test for historicity is reliability. A document that gives general information with vague authorship and audience represents a poor candidate in terms of reliability. Vague content means that the writing could be reproduced years afterward without a strong connection to its original intent.

There are characteristics in a letter that make it more reliable; for example, was the text written: 1) in a personal style (e.g., 2 Corinthians 1); to a specific person or small group (e.g., 1 and 2 Timothy, Philemon, and Titus); 3) with an unpolished hand (e.g., 1 and 2 Peter); or 4) with trivial details (e.g., Mark 14:51–52)?[24]

Another trait of the New Testament is the connection to the original language. The New Testament was written in Koine (classical) Greek. Scholars have full access to the original language, though they do not know its pronunciation. Koine Greek and Modern Greek have many similarities and connections. Compared to other languages, such as English, Greek has changed significantly less over the centuries. Fortunately, a vast amount of classical literature has been written in Koine Greek. Since we have translations of those works into modern languages, we have a language template in Greek. This allows one to make modern translations of the New Testament without the difficulty of leaping through all previous translations and interpretations.[25] There is no such game of Telephone[26] going on here. A Bible scholar need not start with today's English translation and then work through every previous translation to understand the original idea. With all the ancient manuscripts available, scholars have substantial evidence that the NT is valid. In the words of Bart Ehrman:

> Almost all of these variants [of the New Testament] are minor, and most of them are spelling or grammatical errors. Almost all can be explained by some type of unintentional scribal mistake, such as poor eyesight. Very few variants are contested among scholars, and few or none of the contested variants carry any theological significance. Modern biblical translations reflect this scholarly consensus where the variants exist, while the disputed variants are typically noted as such in the translations.[27]

There are even more indications that the New Testament satisfies the historicity test.[28] Historical corroboration of Jesus is perhaps one of the more powerful confirmations that he was an authentic historical person.

Historical corroboration of Jesus

Historical evidence for Jesus of Nazareth is abundant, widespread, and long established. The account of the Nazarene, of course, does not exist in a vacuum, nor is his life only recounted in the Bible. According to the New Testament Scholar, Bart Ehrman, "Virtually all historians and scholars have concluded Jesus did exist as a historical figure."[29]

Detailed biographical accounts of Jesus appeared soon after his death. Eyewitnesses include, of course, the leading Bible writers at the time, Matthew, Peter, John, and James, the brother of Jesus. Though we do not have contemporaneous writings on Jesus, writers wrote about him after a relatively short gap after his death. The earliest writings occurred about 25 years after he died. There are also non-Christian sources, including both Jewish and Roman writers. They both mention Jesus as a historical figure, and these writers were also nearly contemporaneous with the events.[30] James Dunn expresses this recognized historical event: "Jesus was baptized by John the Baptist and was crucified by order of the Roman Prefect Pontius Pilate."[31]

Writings from the Jewish historian Flavius Josephus (37–c. 100 CE), a non-Christian Jew, are compelling. His writing on Jesus recorded in Book 18, Chapter 3, Point 3, which is often referred to as the Testimonium Flavianum (93–94 CE), reads:

> Now there was about this time Jesus, a wise man, if it be lawful to call him a man; for he was a doer of wonderful works, a teacher of such men as receive the truth with pleasure. He drew over to him both many of the Jews and many of the Gentiles. He was [the] Christ. And when Pilate, at the suggestion of the principal men among us, had condemned him to the cross, those that loved him at the first did not forsake him; for he appeared to them alive again the third day; as the divine prophets had foretold these and 10 thousand other wonderful things concerning him. And the tribe of Christians, so named from him, are not extinct at this day.[32]

Two additional citations are given in this account, noting that Jesus was crucified on April 3 and that he reappeared alive on April 5 CE.[33] Josephus indicates here that Jesus could have been the long-awaited Messiah of the Jews. Many suggested that medieval scribes doctored the testimony, but this seems unlikely, as a quote by Agapitos, a tenth-century bishop of Hierapolis, shows that the Arab version of the same text coincides:

> At that time, a wise man called Jesus, admirable in his conduct, was renowned for his virtue. Many Jews and other peoples were his disciples. Pilate condemned him to death by crucifixion. But those who had become his disciples did not renounce their discipleship and told of how he appeared to them alive three days after the crucifixion, and that because of this, he could be the Messiah of whom the prophets had said such marvellous things.[34]

About 20 years later, the Roman politician Tacitus reported that Jesus was executed while Pontius Pilate was the Roman prefect in charge of Judaea (26–36 CE) and Tiberius was emperor (14–37 CE). These reports fit with the timeframe of the Gospels. Tacitus gave a negative view of the Christian religion, calling it destructive. Pliny, a contemporary of Tacitus, was another Roman governor who also disliked Christians, calling their following "pig-headed obstinacy" in their worship of Christ as God.[35]

Other evidence of Jesus comes from writings of the first and second centuries CE that describe how Christians lived. Though the literature was often critical of Christianity, it revealed the existence and the underlying spirit of the brotherhood of the followers. Jesus was denounced as the illegitimate son of Mary and painted as evil. Celsus (178 CE), for example, was a Greek philosopher and opponent of early Christianity, and he dismissed Jesus as a scoundrel. Celsus' disdain toward Christians and Jews is revealed in the tone of his writing, and we see a skepticism that transcends the centuries:

> Jesus was born in adultery and nurtured on the wisdom of Egypt. His assertion of divine dignity is disproved by his poverty and his miserable end. Christians have no standing in the Old Testament prophecies, and their talk of a

resurrection that was only revealed to some of their own adherents is foolishness. Celsus indeed says that the Jews are almost as ridiculous as the foes they attack; the latter said the saviour from Heaven had come, the former still looked for his coming. However, the Jews have the advantage of being an ancient nation with an ancient faith. The idea of an Incarnation of God is absurd; why should the human race think itself so superior to bees, ants and elephants as to be put in this unique relation to its maker? And why should God choose to come to men as a Jew? The Christian idea of a special providence is nonsense, an insult to the deity. Christians are like a council of frogs in a marsh or a synod of worms on a dunghill, croaking and squeaking, 'For our sakes was the world created.' It is much more reasonable to believe that each part of the world has its own special deity; prophets and supernatural messengers had forsooth appeared in more places than one. Besides being bad philosophy based on fictitious history, Christianity is not respectable.[36]

The Christian community, in their reverence to Jesus, was also ridiculed by writers. The Roman satirist Lucian is of particular historical interest because his writing contains one of the earliest evaluations of early Christianity by a non-Christian author.[37]

As a playwright of the second century, we have two quotes by him from a play entitled *The Passing of Peregrinus*. The main character, Peregrinus, was a cynic philosopher who became a Christian, rose in prominence in the Christian community, then returned to cynicism. Here Lucian makes fun of the character of Peregrinus, an individual who took advantage of the Christians and then flipped on his decision, reverting to his previous philosophy. Lucian recounts the words of a speaker in a town square just after a man named Proteus has thrown himself into a fire:

> It was then that he [Proteus] learned the wondrous lore of the Christians, by associating with their priests and scribes in Palestine. And—how else could it be?—in a trice he [Proteus] made them all look like children, for he was [a] prophet, cult-leader, head of the synagogue, and everything, all by himself. He interpreted and explained some of their

books and even composed many, *and they revered him as a god, made use of him as a lawgiver, and set him down as a protector, next after that other, to be sure, [whom]*[38] *they still worship, the man who was crucified in Palestine because he introduced this new cult into the world* [...] So it was then in the case of Peregrinus; much money came to him from them by reason of his imprisonment, and he procured not a little revenue from it.[39] (emphasis added)

Despite the cynicism in the passage, the evidence for Jesus, by a skeptic, is like gold to the scholar. We find another reference to Jesus in the same text:

Then, too, their first lawgiver persuaded them that they are all brothers...after they have thrown over and denied the gods of Greece and have done reverence to that crucified sophist himself and live according to his laws.[40]

Although non-Christian scholars might argue about the details of Jesus' life, the historicity of Jesus' existence is corroborated, perhaps best, by the critics. Beyond the historical evidence is the attestation related to the writings and prophecies of the Bible. We invite the reader to review Appendices 4 and 5, showing the connection of the prophecies from the OT with their fulfillment in the life, death, and resurrection of Jesus.

Chapter summary and questions
One can find substantial evidence for God by exploring the biblical history, archaeology, and prophecy. The center of the Christian faith is the resurrection of Jesus—an event firmly established in history with biblical eyewitnesses along with important confirmations by near-contemporary non-Christian writers.

Questions to be pondered

- Besides the three questions at the beginning of this chapter, can you think of any other questions that a healthy skeptic should pose regarding the Bible? Explain.

- Can you think of any problems archaeological findings might pose in terms of corroborating the Old and New Testaments? If yes, what are they? Why are they problematic? If not, why not?
- Were you ever an eyewitness to an extraordinary event? Did anyone ever doubt you? If yes, what did you do to convince your listener?

NOTES

[1] Thomas L. Thompson, *Biblical Narrative and Palestine's History: Changing Perspectives 2* (London: Routledge, 2014), 164.

[2] J. W. Wallace, *Cold Case Christianity: A Homicide Detective Investigates the Claims of the Gospels* (Colorado Springs: David C. Cook, 2012).

[3] "The Tel Dan Inscription: The First Historical Evidence of King David from the Bible," *The Biblical Archaeology Society*, May 2, 2019 https://www.biblicalarchaeology.org/daily/biblical-artifacts/the-tel-dan-inscription-the-first-historical-evidence-of-the-king-david-bible-story, last accessed, August 2018.

[4] John L. McKenzie, *Dictionary of the Bible* (New York: Simon and Schuster, 1995).

[5] A. H. Layard, *Discoveries Among the Ruins of Nineveh and Babylon* (New York: G.P. Putname, 1853), 12.

[6] A. H. Layard, *The History of Assyria in the Ruins of Nineveh, Royal Asiatic Society* (London: John W. Parker and Son, 1852), 23.

[7] E. Bleibtreu, "Grisly Assyrian Record of Torture and Death," *Biblical Archaeology Review* 17:01 (Jan/Feb 1991).

[8] Stephanie Page, "A Stela of Adad-nirari III and Nergal-eres from Tell al Rimah," *Iraq* 30 (1968): 143.

[9] Several extrabiblical confirmations exist beyond the Assyrian conflicts, including military campaigns and conflicts with Egypt, Moab, Samaria, Babylon, Medeo-Persia, and Rome. Even the Genesis flood event has been corroborated in the Sumerian Gilgamesh Epic. Gary Rendsburg, "The Biblical flood story in the light of the Gilgamesh flood account," in *Gilgamesh and the World of Assyria*, eds. J. Azize and N. Weeks (Leuven, Belgium: Peeters, 2007), 117.

[10] Craig L. Blomberg, *Jesus and the Gospels: An Introduction and Survey* (2nd Edition) (Nashville, TN: B&H Academic, 2009), 424–425.

[11] Ibid.

[12] Ibid.

[13] Paul Rhodes Eddy, and Gregory A. Boyd, *The Jesus Legend: A Case for the Historical Reliability of the Synoptic Jesus Tradition* (Grand Rapids, MI: Baker Academic, 2008), 309–262.

[14] Ibid, 237–308. Also, Craig L. Blomberg, *Historical Reliability of the Gospels* (Downers Grove, IL: InterVarsity, 1986), 19–72.

[15] Blomberg, *Jesus and the Gospels: An Introduction and Survey*, 424.

[16] A few include: E. P. Sanders, *The Historical Figure of Jesus* (New York: Penguin, 1993), Craig A. Evans, "Life-of-Jesus Research and the Eclipse of Mythology," *Theological Studies* 54 (1993), 5, and R. M. Grant, *A Historical Introduction to the New Testament* (New York: Harper and Row, 1963).

[17] Two sources are: Dominic Crossan and Richard G. Watts, *Who Is Jesus? Answers to Your Questions about the Historical Jesus* (Louisville, KY: Westminster John Knox, 1999), 108, and Bruce M. Metzger, *Textual Commentary on the Greek New Testament*, United Bible Societies Greek New Testament, 4th ed. (2006), where he notes that Luke 24:51 is missing in some important early witnesses, and Acts 1 varies between the Alexandrian and Western versions.

[18] The earliest manuscript is a business-card-sized fragment from the Gospel of John, Rylands Library Papyrus P52 (Manchester, UK).

[19] Bart D. Ehrman, *The New Testament: A Historical Introduction to the Early Christian Writings* (New York: Oxford University Press, 2008), 193.

[20] Josh McDowell, *Evidence That Demands a Verdict*, rev. ed. (San Bernardino, CA: Here's Life, 1979), 42.

[21] Updated Manuscript comparison by Josh McDowell (August 8, 2014), https://www.josh.org/wp-content/uploads/Bibliographical-Test-Update-08.13.14.pdf

[22] One close exception (but still four times the time gap) is Thucydides, History of the Peloponnesian War (ca. 460–404 BCE), with papyri fragments existing from the third century BCE. This produces a gap of only 200 years. The Leuven Database of Ancient Books lists 96 papyrus and parchment manuscripts, http://www.trismegistos.org/ldab/search.php. Accessed 2/2/2011. See also, Thucydides, *The Peloponnesian War*, P. J. Rhodes, trans. (New York: Oxford University Press, 2009), 663.

[23] See Norman L. Geisler and William E. Nix, *A General Introduction to the Bible* (Chicago: Moody, 1968), and Bruce M. Metzger, *The Text of the New Testament: Its Transmission, Corruption, and Restoration* (New York: Oxford University Press, 1964). Also, slightly more recently, Gordon D. Fee, "The Textual Criticism of the New Testament," in Introductory Articles, Vol.1 of *The Expositor's Bible Commentary*, ed. Frank E. Gaebelein (Grand Rapids, MI: Zondervan, 1979), 419–33.

[24] G. Louis, *Understanding History: A Primer of Historical Method,* 2nd ed (New York: Alfred A. Knopf, 1969), 53–54. For a good discussion of this topic, see G. L. Bahnsen, "The Inerrancy of the Autographs" in *Inerrancy,* ed. Norman L. Geisler (Grand Rapids, MI: Zondervan,1980), 151–93.

[25] This is a big misconception concerning translations. If the original translation is available, the current translation can be compared directly.

[26] Telephone is a popular party game where the first person in a line whispers a message to the second player, then the second player whispers to the third, and so on, until the message reaches the end of the line, where it's often unrecognizable from the original message.

[27] Bart D. Ehrman, *Misquoting Jesus: The Story Behind Who Changed the Bible and Why* (San Francisco: Harper, 2005), 88–89.

[28] Ehrman, *Misquoting Jesus*, 59.
[29] Bart D. Ehrman, *Did Jesus Exist? The Historical Argument for Jesus of Nazareth* (New York: HarperOne, 2012), Chapter 1.
[30] In contrast, there are orders-of-magnitude longer periods of time ascribed to mythical figures such as King Arthur, who supposedly existed around the fifth century CE, but written accounts occurred only in the ninth century CE. S. Gathercole, "What is the historical evidence that Jesus Christ lived and died?" (Apr 2017), www.theguardian.com, accessed August 2018. https://www.theguardian.com/world/2017/apr/14/what-is-the-historical-evidence-that-jesus-christ-lived-and-died
[31] James D. G. Dunn, *Jesus Remembered* (Grand Rapids, MI: Eerdmans, 2003), 779–781.
[32] Flavius Josephus, "The Testimonium Flavianum," *Antiquities of the Jews*, Book 18, Chapter 3, point 3 (circa 71 CE), based on the translation of Louis H. Feldman, The Loeb Classical Library, http://www.josephus.org/testimonium.htm.
[33] Another proposed date is his death on 7 April 30 CE. Colin J. Humphreys and W. G. Waddington, , "The Date of the Crucifixion," *Journal of the American Scientific Affiliation,* March 1985, 37.
[34] Opus Dei, Christian life: What do Roman and Jewish sources tell us about Jesus? https://opusdei.org/en-us/article/what-do-roman-and-jewish-sources-tell-us-about-jesus/, last accessed 2 October, 2018.
[35] Gathercole, "What is the historical evidence that Jesus Christ lived and died?"
[36] "Celsus, The True Word" (or Account, ἀληθὴς λόγος), was published by Origen in 248 CE, 70 years after its composition. Ed. Hugh Chisholm, "Celsus," *Encyclopædia Britannica,* 11th ed. (Cambridge University Press, 1911).
[37] Robert E. Van Voorst, *Jesus Outside the New Testament: An Introduction to the Ancient Evidence* (Grand Rapids, MI: Eerdmans, 2000), 58–64.
[38] More directly from the original text: "protector; that great man, to be sure, they still worship,"
[39] Lucian, "The Death of Peregrine" sections 11–13 (circa 165–175 CE), trans. H.W. Fowler and F.G. Fowler, *The Works of Lucian of Samosata* vol. 4, (Oxford: Clarendon Press, 1949), as quoted in Gary R. Habermas, *The Historical Jesus: Ancient Evidence for the Life of Christ* (Joplin, MO: College Press, 1996).
[40] Ibid, Section 13.

Chapter 9 – God of the Gaps or Gaps about God?

Tell people there's an invisible man in the sky who created the universe, and the vast majority will believe you. Tell them the paint is wet, and they have to touch it to be sure. – George Carlin

Though I do not agree with everything George Carlin said in his comedy career, including this common misconception of God, I do agree with his observation about wet paint. The "science of finding God" has many layers, but ultimately, it comes down to touching wet paint, so to speak.

Science and faith are indeed connected. We make progress in each by exploring, watching, touching, measuring, and testing. The path in science, though far more technical, is in practice simpler. The scientific method is all about following a careful set of steps and being meticulous as we try to understand the world around us, whether it be stars, planets, rocks, plants, zebras, or lionfish. If scientists make mistakes, they get corrected in the next scientific inquiry. Egos may be hurt if someone's publication is not accepted, a research grant is not awarded, or perhaps a paper is rejected. But at the end of the day, science makes progress, and we all benefit from this discipline.

On the other hand, faith is more complicated and messier. Instead of test tubes, telescopes, and T squares, faith requires learning by watching others. But when done wholeheartedly, this is a beautiful way to grow in faith and ultimately a way to learn how to imitate Jesus (1 Corinthians 11:1). But when done halfheartedly or simply incorrectly, faith becomes more of a system of uninspiring rules, an empty religion, or even a perversion—and skeptics, understandably, are not impressed.

Christianity is marked with bad practice and has had periods during which believers either misunderstood the message, misinterpreted the doctrine, practiced the doctrine in a very unsatisfactory way, or willfully ignored it for personal gain. But once we remove the dirty layer, it's remarkable what we can find. This chapter is all about removing that layer.

Rebooting our thinking

Those who work with computers but are not necessarily knowledgeable with the details of them know about the power of the reboot. When software is frozen, or bizarre behavior materializes on your machine, the easiest way to move forward is to address the on-off switch. We can say the same about the way we view the world and the way we see God. Rebooting is helpful for everyone, including the dogmatic atheist, the fundamentalist, or just inquisitive Jane or Joe.

A trained scientist generally views the world with logic and clarity (at least in their domain of expertise). But if a scientist has a preconceived notion about nature, they are asked to leave their notions aside. Scientists perform tests, and if their view is proven wrong, they must get past their error, change direction if needed, and move forward. The reason this works reasonably well in science (although there are notable exceptions!) is that a scientist's goal is to seek the truth on a matter. We suggest a similar mindset when we think about God. Preconceived notions are powerful inhibitors, and the only way to move forward to better understanding God is to examine the evidence with an open mind.

Years ago, I went through this exercise as I was learning about God. I had to do some "reboots" of my thinking, and I continue to reboot from time to time when I get off track. In my journey to understand God, I carried over some thinking that was rooted in naturalism. I needed to be open to the fact that not all of God's characteristics fit neatly into a scientific explanation, nor should they. At the same time, I had a few concepts in my head that stemmed from traditional religion, residual thoughts from my early Christian education. As I reread the Bible in my twenties, I started to pay attention to context and personalized lessons. Though this step was not easy, I cannot overemphasize the parallel with a scientific exercise. I attempted to root out issues that cause unclarity (sin) and emphasize the spiritual teachings of Jesus that clarify. Just as science ultimately reveres those who dare say they were wrong and rethink things, the Bible's message lauds those who do the same. Though this reboot (repentance) was difficult, I understood it was essential to take in Jesus' message.

To the unknown god

I once had the opportunity to go to a scientific conference in Rhodes, Greece. My family joined me on this trip, and one free evening, we visited the ancient city and the port. Centuries ago, there stood at the entrance of this port an enormous statue of Helios, god and personification of the Sun. This statue was known as the Colossus of Rhodes. It was constructed in 280 BCE, measured 33 meters high, and stood on a pedestal of 15 meters (altogether, similar in height to the Statue of Liberty without its foundation). Helios held a torch and acted as a lighthouse for approaching ships; he was a wonder of the ancient world, and the ancient Greeks adored him. His statue was perhaps the most famous one in ancient Greece, but it was only one among thousands. In 226 BCE, an earthquake toppled Helios, and no one ever rebuilt it.

Few today (I would hope!) would argue that a god lived in that statue and that a "spirit" surveyed the ancient port and protected its ships. If the Colossus of Rhodes stood today, we might be impressed by the structure and beauty of the artwork, but nothing beyond that.

This statue is an example of George Carlin's "little man in the sky," and the idol was indeed brought down to the Earth in a dramatic way through the forces of nature.

A few hundred miles from Rhodes, in the city of Athens, the Apostle Paul reflected on idols and deity. The year was 50 CE. The Athenians of the first century, those who were both academically and religiously inclined, loved to discuss topics of this nature. But concerning God, the Athenians needed a mental reboot:

> While Paul was waiting for them in Athens, he was deeply distressed to see that the city was full of idols. So he argued in the synagogue with the Jews and the devout persons, and also in the marketplace every day with those who happened to be there. Also some Epicurean and Stoic

philosophers debated with him. Some said, "What does this babbler want to say?" Others said, "He seems to be a proclaimer of foreign divinities.p" (This was because he was telling the good news about Jesus and the resurrection.) So they took him and brought him to the Areopagus and asked him, "May we know what this new teaching is that you are presenting? It sounds rather strange to us, so we would like to know what it means." Now all the Athenians and the foreigners living there would spend their time in nothing but telling or hearing something new. (Acts 17:16–21)

Intellectuals, philosophers, thinkers, and religious people made up the gathering at the Areopagus. These people were neither unintelligent nor ignorant; some onlookers were exposed to the best science of the day, and though incomplete, it was still sophisticated. They wondered about Paul's God, his "little man in the sky." How would he be fed? How could people meet his needs? He would surely need assistance. But they all got it wrong—completely wrong. The atmosphere of the Areopagus was filled with confusion and misconceptions about God. This scene reminds me of the conversations in the coffee house I once frequented on my campus. Topics were all over the place (philosophy, religion, math, Star Trek), and most of them were way off in terms of importance, or just way out there, in a place no one had gone before. But I admit it was fun being part of these conversations.

But the Bible spends no time trying to convince people of God's existence; rather, it assumes that God exists and that people are (or were) all aware. People are not the originators of God; it's just the opposite. But we have replaced God with religion, and Paul saw this.

> Then Paul stood in front of the Areopagus and said, "Athenians, I see how extremely religious you are in every way. For as I went through the city and looked carefully at the objects of your worship, I found among them an altar with the inscription, 'To an unknown god.' What therefore you worship as unknown, this I proclaim to you." (Acts 17:22–23)

When we read this, it's easy to displace our concern. Paul is correcting the Athenians, but is this message is entirely inappropriate for us today? If we listen carefully, we might find something for us.

We add an extra layer of sophistication to our modern gods; though we might not consider magic, the supernatural, or the mysterious, we might be prone to place important items on our favorite pedestal. The adoration of our lifestyle, personal possessions, relationships, and self-worth might rival the Greeks' veneration of their goddess Hedone.[1] Our prioritizing of work, title, and power would put the Greeks' worship of Hephaestus[2] to shame. The way we have elevated the philosophy of naturalism is probably not distinguishable from the way people esteemed the goddess Athena.[3] Even religious people can slip into the Areopagus and unwittingly worship Zeus;[4] we do this when we allow ourselves to feel better or superior to others because we follow God. But our gods are just as empty, hard, and rigid as any of the gods in the hallways of the Pantheon.

Not only have we metaphorically resurrected the Greek gods for the twenty-first century, but we have also become comfortable with them—they are perfectly designed to fit our culture and our lifestyle. But the Greeks were far from doltish; they were also savvy enough to create a god for all their needs. They even figured out a way to create a god of indecisiveness. They labeled his altar "TO AN UNKNOWN GOD." He goes by another name, *Agnostos Theos*.[5]

When we think we have all the answers, either scientifically, philosophically, religiously, or all three, and we are no longer learners, we end up scoffing. We will not tolerate the inconvenience when someone like a Paul enters our midst, especially when he comes with news about Jesus. We become less flexible in our thinking. But proper science and good religion both require regular effort. We should step back, see the big picture, and rethink our conviction using the evidence that is accessible to us. We might find it useful to listen to Paul as he describes God:

> "The God who made the world and everything in it, he who is Lord of heaven and earth, does not live in shrines made by human hands, nor is he served by human hands, as

though he needed anything, since he himself gives to all mortals life and breath and all things." (Acts 17:24–25)

If you are struggling with Paul's words, you are not the only one. For many years, I identified with the Athenians—they scoffed at Paul as soon as he started talking about the miraculous powers of Jesus (Acts 17:32a). For many years, I too was a scoffer. I inwardly laughed when I saw a Christian pray or sing. But not everyone scoffs. Some of those Athenians did invite Paul back (Acts 17:32b) to listen to what he had to say—eventually, I did the same.

God of the gaps? Really?
Victor Stenger is a particle physicist and an advocate of philosophical naturalism, skepticism, and atheism. He vigorously and elegantly argues against the concept of a "God of the gaps," or for short, GOG. According to Stenger, GOG is unnecessary, since science will eventually explain everything.[6] I am going to write something that may shock some Christians: I partially agree with Dr. Stenger. But to clarify, the concept of GOG does not represent the God of the Bible.

The term "God of the gaps" is a theological argument to describe the gaps or holes in scientific knowledge as evidence or proof of God's existence. Critics such as Stenger understood this phrase to mean that God is a bogus explanation for anything not currently explained by science. Theologians initially used this argument to support the existence of God, taking as evidence that science has no explanation for complex or unknown phenomena.

The phrase dates back to a lecture by evangelist Henry Drummond. Drummond corrected his Christian audience, who pointed to things in nature as evidence for God. Instead, he urged people to believe in the God of all nature and not just an "occasional wonder-worker, who is the God of an old theology."[7]

In this book, we have attempted to turn GOG on its head, because it's not the scientific gaps that give us support for God; it's what we observe that provides evidence for God. Other scientists today, such as Francis Collins, also reject GOG while embracing "the idea of a God who fine-tuned the universe precisely so that life could exist."[8] German theologian and martyr Dietrich Bonhoeffer eloquently expressed the concept in similar terms in letters he wrote while in a Nazi prison.

> How wrong it is to use God as a stop-gap for the incompleteness of our knowledge. If in fact the frontiers of knowledge are being pushed further and further back (and that is bound to be the case), then God is being pushed back with them, and is therefore continually in retreat. We are to find God in what we know, not in what we do not know.[9]

As science pushes forward, the wonders of creation provide us with an easy way to appreciate the creator, a level of appreciation only humans can obtain. Bonhoeffer's premise is on this point. It's also true that some gaps become wider over time. While science reveals the workings of the surface layer of nature, it also leads us to irreconcilable hypotheses when no further gain in knowledge will fill the gaps. In the author's opinion, gaps in scientific knowledge should not be a place where we insert a deity. Instead, a gap is a place where we can congregate, have fruitful discussions, and appreciate the complexities of the universe.[10]

Illogic does not dismiss God

Have you ever been asked a question that had no answer? Or has someone ever asked you a question and was seeking obscurity, not clarity? These types of questions are raised about God all the time (often in the scientific community). The fact that sometimes there are no answers to them does not mean that God does not exist. Perhaps the question was poorly posed. I know some who get wrapped up in philosophy or imagine impossible situations in their minds, and then leap to the conclusion that God cannot exist. The number 1 cannot be equal to the number 2: $1 \neq 2$. Someone might decide that $1 = 2$ and immediately make the extension that God cannot exist. In other words, someone can construct a wrong equation or make an incorrect statement and remove God's existence all in the same sentence. I always wondered how that worked.

There is a related example that goes as follows: "Can God create a rock so heavy that he cannot lift it?" If one answers yes, then God is somehow weak since he cannot lift the rock, but if he can lift it, his powers are again limited, since he could not create a rock heavy enough; therefore, he is not almighty.

The point here is a misunderstanding in the definition. The quality of being all-powerful does not mean that God's power can overcome illogic and therefore perform illogical, contradictory acts. As humans, we can conjure up just about any definition that we wish and slap flawed concepts together. Our creative and sometimes unreasonable behavior speaks more to the wonderment of free and abstract thinking than to the flaws of a Creator.

An omnipotent God should be able to do anything *logical*. Miracles, therefore, should not be outside the realm of God's power. God should be able to create something from nothing. He should be able to make something alive that is not alive. The Bible is replete with examples of miracles, but the Bible never indicates that God can make 1 = 2, square circles, four-sided hexagons, or lines with only one end. God should be able to make a man walk on water, but he should not be able to make a man walk on water and simultaneously not be able to walk on water. But this does not mean God is subservient to logic. According to writer Jim Boucher,

> This is not to say that logic is some sort of force that transcends God or to which he is a slave. Rather, it says that logical consistency is founded in the person of God himself.[11]

Free will is inherently human and central to the nature of God and his relationship to us. This freedom implies that we not only have the choice to make selfish decisions; we also have the freedom to construct faulty logic. None of our crazy ideas or poorly constructed equations diminishes God in the least.

This type of questioning appears in the Bible. More than once, religious leaders tried to trap Jesus in his own words so that he would be found guilty. In the following instance, the religious leaders, instead of going to him to ask these questions (I wonder why), sent others:

> So they asked him, "Teacher, we know that you are right in what you say and teach, and you show deference to no one, but teach the way of God in accordance with truth.

Is it lawful for us to pay taxes to the emperor, or not?" But he perceived their craftiness and said to them, "Show me a denarius. Whose head and whose title does it bear?" They said, "The emperor's." He said to them, "Then give to the emperor the things that are the emperor's, and to God the things that are God's." And they were not able in the presence of the people to trap him by what he said; and being amazed by his answer, they became silent. (Luke 20:21–26)

Had Jesus said, pay taxes, he would position himself with Rome. The Jewish people hated their conquerors, and therefore, this would have turned the Jews against Jesus. Answering no would be akin to treason toward Rome. Jesus was seemingly trapped, but he found a way to step out of this tricky spot and silence his opponents.

Most arguments against God arise from illogical constructions (informal fallacies). The one many of us are familiar with is: "Who created God?" This question supposes that God needed to be created, and the conclusion is already assumed. We should immediately throw out this argument. There are many others like it, and they are all similar. C.S. Lewis once said, "Nonsense is still nonsense even when we speak it about God."[12]

Imperfect practitioners: Hypocrites

I have heard it said that God could not exist because there are many hypocrites in the church. In the past, I have been sympathetic to those who make this claim. Understandably, hypocrisy in religion does not inspire people. I, too, have been guilty of considerable hypocrisy, and I am not proud of it. Furthermore, I have been a hypocrite while calling others hypocrites. I have viewed other religious people with suspicion while I was pursuing a selfish, self-focused life. There are two reasons for the aversion to hypocrisy. One, people have no desire to become what they are seeing in the hypocrite. Two, people doubt God if they see only followers who follow superficially. But if we excuse ourselves for not pursuing God because of imperfect practitioners, we are going down the wrong path.

The unfortunate characteristic of hypocrisy occurs when we know what we ought to do, but don't do it. Sometimes, it's easier to say one thing or act one way in a certain situation and then act another way in a different setting. While in college, I went to church on Sunday; then, in the evening, I secretly rewired the stereo of a resident in my dorm so that his music would play at 3:00 am. My short-lived convictions to be kind to my neighbor applied in church, but when I felt annoyed by my neighbor (who enjoyed classical music), I threw my religious beliefs to the wind and gave him a dose of Bach in the middle of the night.

Hypocrites are actors. The term "hypocrisy" has a root in the Greek *hupokrisis*, "acting of a theatrical part." The Bible points out that hypocrisy is wrong, and none of us likes to be called a hypocrite. But all of us, to one degree or another, are. We present ourselves as law-abiding citizens, but who has not inched over the speed limit, just a bit? Who has not cut someone off on the road and smiled while pretending not to see the person? Who has said, "It's great to see you," yet would rather be someplace else? Who wants to save the planet from climate change but is perfectly content driving an SUV and producing an environmental footprint as large as an entire village in the developing world? No matter what our religious background, if we are honest with ourselves, we might notice the H-label clearly pinned on our lapel.

Hypocrisy is considered a sin. Jesus and NT writers had more to say about this flaw than any other (Matthew 23:28; Mark 12:15; Luke 12:1; Galatians 2:13; 1 Peter 2:1), and Jesus primarily addressed religious leaders, the Pharisees. A poignant passage concerning this topic is: "Anyone, then, who knows the right thing to do and fails to do it, commits sin" (James 4:17).

The fact of the matter is this: combining free will with religion will produce at least some hypocrisy. But even if hypocrisy filled the world, God would not be reduced in the least. As a parallel, imagine the world filled with wacky scientists performing pseudoscience for nefarious gain; even in this case they would not diminish one iota the power of science, only the deficiency of the scientists. Of course, wacky scientists would make it harder for others to understand science and apply it. Likewise, religious hypocrites, though troublesome, should not hinder us from seeking God if we are really serious about it.

Religious confusion

Many are turned off by God because of religious confusion. This confusion is a huge challenge, and I have not forgotten about this "woolly mammoth" in the room. I am not a theologian, so I have looked at this issue through the eyes of a scientist.

In the mid-1980s, I took a special trip to Marseille, France, where I met a physicist named Charles Nahon. Dr. Nahon presented me with a copy of one of his many volumes on the topic of debunking special relativity; in particular, Nahon postulated that the speed of light was constant.[13] He spent his life working on this theory that showed, as he maintained, "Einstein's theory was wrong." It was difficult to follow his work, but ultimately his fundamental point was an ideological one. I carefully considered Charles Nahon's reasoning and read through his work again and again. I still have his book on my office shelf.

Why was I interested in knowing this physicist? The reason was simple: if his work stood the test of the scientific method, he would have entirely revolutionized modern physics. As it was, his theories were incorrect, and he remains one of the many obscure scientists who have attempted to find an alternative theory to relativity. But the only way one could have tested and ultimately discerned between Einstein's theory and Dr. Nahon's was to measure—and this is what science has done.

Scientists following Einstein have tested and retested his postulates, confirming that the speed of light is constant. Though Dr. Nahon's proposition was wrong, I hold no grudge toward him. I don't believe he wasted my time, nor did I resent going to Marseille to meet up with him. I admire his persistence and energy, and if I learned anything from the experience, it taught me much about our topic at hand.

I invested significant amounts of energy in working through Dr. Nahon's hypothesis. It took time, traveling money, reading effort, and some persistence. I had to ask lots of questions and got to know Dr. Nahon. This entire experience, traveling to Marseille to meet this scientist and read his works, allowed me to think about relativity in a way I never thought about it before and in a way that classroom experience never provided. There is something about putting physical and emotional energy as well as financial backing into a pursuit that brings you to appreciate a goal, even if that goal seems outrageous. I thought at the time that my effort could help change the world. The world didn't change, but something inside me clicked.

We could say the same about religion and our pursuit of God. Some pursue God through the systematic study of theology. This rigorous discipline helps us understand doctrines and practices thoroughly. This is one approach that may help the scientifically inclined, but it's certainly not for everyone. Another method is more organic, personal, and perhaps a bit more haphazard. It requires a personal effort to read the Bible, asking questions of others who are doing the same and praying to God for guidance. This approach might seem more random than pursuing a theology degree, but it has its advantages. Anyone can do it at any place and at any time. Furthermore, the Bible points out that God is not an object for us to find; he is a person for us to discover. Both paths take effort, but when it comes to building friendships, there are advantages to the latter approach.

One may find a friend by pursuing all sorts of activities: talking, visiting people, trying to understand people better (or, spiritually speaking, praying, attending church, and reading the Bible). The more determined we are in this endeavor, chances are, the faster we will reach our goal in making a friend. Contrarily, being selfish and self-focused, and showing little or no interest in other people will probably result in few or no friends. The point is that you must at least make an effort to seek God, even if a theology degree is not your intent.

I have not forgotten a significant consideration in terms of God: in some religions, God is not considered a person. Others see God as a person, but he is not approachable as a friend. Though different religious denominations point to the same God, they claim different identities for God—this can't be so. God cannot have opposing identities. He cannot be a personable God and simultaneously an impersonal God. We need to make an effort to understand his true identity, or we risk worshiping something or someone in the wrong way. We cannot take a trip to Marseille to meet God to do this, but we can read about God to understand his character.

This book is not about comparative religion, but a serious student conducting an exhaustive study of religion could come to a clear conclusion on what is and isn't worthy of following. There were 4200 religions in 2019, according to the World Religions Statistics survey, and there are seven major ones: Christianity, Islam, Judaism, Hinduism, Buddhism, Sikhism, and Animism. As with any academic study, an earnest student could examine all these religions, perhaps putting each under a "spiritual microscope." Some people who believe there is one God may need to go through this exercise if they are harboring doubts in their minds.

But if God is all about a relationship, then Paul shows us how to wade through complex religious questions and target the most critical issues in that relationship. Before Paul arrived in Athens, in Thessalonica, he and Silas did what they always did when visiting a city: they talked to others about Jesus.

> And Paul went in, as was his custom, and on three sabbath days argued with them from the scriptures, explaining and proving that it was necessary for the Messiah to suffer and to rise from the dead, and saying, "This is the Messiah, Jesus whom I am proclaiming to you." Some of them were persuaded and joined Paul and Silas, as did a great many of the devout Greeks and not a few of the leading women. (Acts 17:2–4)

But not everyone was interested in the message about Jesus' life. Here is what happened:

> But the Jews became jealous, and with the help of some ruffians in the marketplaces they formed a mob and set the city in an uproar. While they were searching for Paul and Silas to bring them out to the assembly, they attacked Jason's house. When they could not find them, they dragged Jason and some believers before the city authorities, shouting, "These people who have been turning the world upside down have come here also, and Jason has entertained them as guests. They are all acting contrary to the decrees of the emperor, saying that there is another king named Jesus." The people and the city officials were disturbed when they heard this, and after they had taken bail from Jason and the others, they let them go. (Acts 17:5–9)

No doubt, this was a religious mess, and in this case, religious confusion started with the leaders. The Jewish leaders were upset and jealous of this new sect called Christianity. First-century Christianity was bumping up against 18-centuries-old Judaism. For someone not versed in these religions, this may seem like an internal conflict, based on opinions about God. But here we can learn about a universal concept to apply when facing religious confusion.

> That very night the believers sent Paul and Silas off to Berea; and when they arrived, they went to the Jewish synagogue. These Jews were more receptive than those in Thessalonica, for they welcomed the message very eagerly and examined the scriptures every day to see whether these things were so. Many of them therefore believed, including not a few Greek women and men of high standing. (Acts 17:10–12)

Paul's and Silas' reaction to the mob scene in Thessalonica was to go elsewhere. Berea was 72 kilometers or about 45 miles down the road, and people there were more open to listening and reasoning. The NIV translation of the above passage reads that "the Berean Jews were of more noble character than those in Thessalonica." The New Living Translation uses the term "open-minded," and the New American Bible translation uses the term "fair-minded."

This small community in northern Greece found the key to addressing religious confusion, and there is no reason to believe that the Berean practice could not be applied anywhere and to any century. They had a genuine curiosity about God and were determined to understand what Paul was teaching. In a relatively short amount of time, they answered their own questions by going back and reading Scripture. Using their same technique, we could perhaps untangle all 4200 religions and distill that which is essential in building a relationship with God.

Fortunately for us, the effort needed to build a friendship with God is less complicated than trying to discern between good and bad theories of relativity. But for someone serious about their research of God, I suggest this: the effort and passion needed to pursue God should be no less than any other effort or pursuit we undertake.

For the Bereans, there was both an intellectual and an emotional interest in God, and it involved not just a single sitting with the Bible, "for they welcomed the message very eagerly and examined the

Scriptures every day to see whether these things were so." From my experience, if one does not take time to read the Scriptures with a certain open-mindedness, learning not just intellectually but with a degree of emotion, and does not persevere, Christianity will forever be one more confused religion among many others. Could we not say the same about any academic pursuit on any topic?

Paul's visit to the Bereans was in 50 CE or about 20 years after Jesus' crucifixion. The Scriptures the Bereans were reading were dated from hundreds to more than a thousand years earlier, in what today is the compilation of the Old Testament. The Bereans were searching for the fulfillment of predictions or prophetic text in the life of Jesus, and they found it. One might even say that the Bereans were scientific in their inquiry: they had a hypothesis and were testing it.

Berean-like qualities are rare in the Christian community; I suspect that they are rare in other religious communities as well. Maybe one out of a hundred Christians I encounter spends even a few minutes reading their Bible monthly, let alone daily. Most follow their religion because it's convenient or because they inherited their beliefs from their family. But religious confusion can be reduced to a minimum if we do what the Bereans did: read the Bible and ask questions. I liken this to a scientist, an enthused one to be sure, embarking on a scientific quest. That scientist would never persevere unless he had a hypothesis that he was going to find something worth his effort. What is remarkable in Christianity is that most know what the reward is, yet spend precious little time in the search.

Is God hiding?

Referencing the preface of this book, in the early 1980s, while I was still trying to figure out my spiritual path, a fellow physics student and atheist (Serge) threw out the challenge, "If God is real, let him make himself appear right in front of us now, and I will believe." As you recall from the preface, Serge was not impressed and I was embarrassed—I was still learning.

Over the years, I reflected on this "noneventful" moment. But God was not hiding; I simply had a misconception of him. He is never obligated to perform a show for us and certainly not for two unspiritual college students. Nonevents were the norm for the Israelites. For centuries God did not appear to his chosen people in the period between the OT and NT. Even the prophet Anna, who prayed day and night in the temple, waited until the end of her life to see the Lord, and when he appeared, he was just a baby (Luke 2:36–38).

Even if God did suddenly show up in our dorm, in front of Serge and me, I am wondering what we would have done. We might have been stunned, shocked, and shivering. Perhaps we would have been motivated to be "good" people, at least for a few days or so. But the feeling would have eventually worn away, and both of us would probably have continued our lives with little or no change.

The people in Jesus' hometown were genuine witnesses to many miracles, but even that did not produce faith.

> He left that place and came to his hometown, and his disciples followed him. On the sabbath he began to teach in the synagogue, and many who heard him were astounded. They said, "Where did this man get all this? What is this wisdom that has been given to him? What deeds of power are being done by his hands! Is not this the carpenter, the son of Mary and brother of James and Joses and Judas and Simon, and are not his sisters here with us?" And they took offense at him. Then Jesus said to them, "Prophets are not without honor, except in their hometown, and among their own kin, and in their own house." And he could do no deed of power there, except that he laid his hands on a few sick people and cured them. And he was amazed at their unbelief. (Mark 6:1–6a)

If you are not genuinely seeking God, a miracle will produce a short-lived "Wow!" According to this passage, the only one who would be genuinely amazed is Jesus—reflecting on just how little faith we have.

Skeptics claim that God is hiding because he does not exist; this was the position of my physics colleague Serge. We could both go to the lab and examine an electron; we could build a machine that measures its mass and electric charge: the electron cannot hide. We could go to the observatory and measure a planet's eccentric orbit: we measure it, end of story. Some particles are challenging to measure, and some heavenly objects are difficult to detect, but they ultimately cannot hide. In terms of hiding, C.S. Lewis made an insightful comment when it came to animals, humans, and God:

> But suppose you are a zoologist and want to take photos of wild animals in their native haunts.... The wild animals will not come to you; but they can run away from you. Unless you keep very quiet, they will. There is beginning to be a tiny little trace of initiative on their side. Now a stage higher; suppose you want to get to know a human person. If he is determined not to let you, you will not get to know him. You have to win his confidence. In this case the initiative is equally divided—it takes two to make a friendship. When you come to knowing God, the initiative lies on His side. If He does not show Himself, nothing you can do will enable you to find Him. And, in fact, He shows much more of Himself to some people than to others—not because He has favorites, but because it is impossible for Him to show Himself to a man whose whole mind and character are in the wrong condition.[14]

Jesus also told a story about God who "hides": the story about the rich man and a poor beggar named Lazarus. Lazarus once fed himself on crumbs that fell from the rich man's table. Both men have died:

> "The poor man died and was carried away by the angels to be with Abraham. The rich man also died and was buried. In Hades, where he was being tormented, he looked up and saw Abraham far away with Lazarus by his side. He called out, 'Father Abraham, have mercy on me, and send Lazarus to dip the tip of his finger in water and cool my tongue; for I am in agony in these flames.' But Abraham said, 'Child, remember that during your lifetime you received your good things, and Lazarus in like manner evil things; but now he is comforted here, and you are in agony. Besides all this, between you and us a great chasm has been fixed, so that those who might want to pass from here to you cannot do so, and no one can cross from there to us.'" (Luke 16:22–26)

This troubling scene reveals something about the heart of a person. When spiritual realities become apparent, we can quickly redirect our priorities and change our demeanor: we can repent. The rich man, once living in comfort and oblivious to the needs around him, had a complete change of heart once he had died. He even turned evangelistic.

> "He said, 'Then, father, I beg you to send him to my father's house—for I have five brothers—that he may warn them, so that they will not also come into this place of torment.' Abraham replied, 'They have Moses and the prophets; they should listen to them.' He said, 'No, father Abraham; but if someone goes to them from the dead, they will repent.' He said to him, 'If they do not listen to Moses and the prophets, neither will they be convinced even if someone rises from the dead.'" (Luke 16:27–31)

Miracles cannot convince the inconvincible. Miracles were done throughout the Bible, many directly in front of skeptics, but never did a miracle produce faith on the part of an unwilling audience. I suppose if Jesus appeared every century, or even every year, and performed a few miracles, the result would not be hundreds of millions, or billions, of people believing and sincerely following him. The events would probably turn into a circus spectacle, which people would attempt to commercialize for some monetary gain.

As a side note, it's interesting that Jesus named a specific person in the story: Lazarus. The rich man was not named. Some have suggested that the rich man was a representative person. Perhaps he characterizes many of us, oblivious to the needs around us. Jesus responded in a straightforward way to this attitude:

> When the crowds were increasing, he began to say, "This generation is an evil generation; it asks for a sign, but no sign will be given to it except the sign of Jonah. For just as Jonah became a sign to the people of Nineveh, so the Son of Man will be to this generation. The queen of the South will rise at the judgment with the people of this generation and condemn them, because she came from the ends of the earth to listen to the wisdom of Solomon, and see, something greater than Solomon is here! The people of Nineveh will rise up at the judgment with this generation and condemn it, because they repented at the proclamation

of Jonah, and see, something greater than Jonah is here!" (Luke 11:29–32)

Jesus' response to those seeking a miracle is that they should look at "the sign of Jonah"—the resurrection. The Apostle Paul did not take this point lightly.

> For I handed on to you as of first importance what I in turn had received: that Christ died for our sins in accordance with the scriptures, and that he was buried, and that he was raised on the third day in accordance with the scriptures, and that he appeared to Cephas, then to the twelve. Then he appeared to more than five hundred brothers and sisters at one time, most of whom are still alive, though some have died. Then he appeared to James, then to all the apostles. Last of all, as to one untimely born, he appeared also to me. (1 Corinthians 15:3–8)

Apart from John, it's possible that most, perhaps all, of the apostles died as martyrs (see Appendix 5); had they denied the resurrection, they might have all lived to see another day—but they did not. Though we may not expect Jesus to show up at our whimsical beckoning today, the evidence we have of his resurrection should get our attention.

A story woven through the fabric of time

The Bible is all about relationships; the primary relationship it describes is between God and humankind. This relationship story threads its way through more than 21 centuries from Abraham to Jesus, and there are countless other centuries between the "creation week" and Abraham. During this expanse of time, culture, human practices, and desires to seek God have waxed and waned. Though God has never changed, humankind's fickle relationship with him changes all the time. Expectations in that relationship also had to evolve; later generations were supposed to learn from earlier generations. As a child grows and has more responsibility, God also expected humanity to learn lessons from those who went before. Jesus filled his Sermon on the Mount with teaching that called the current generation above and beyond the spirituality of earlier generations (e.g., Matthew 5–7).

Relationships are also complex. They are sometimes illogical and ironic, and unfortunately, they are occasionally tragic. The Bible was not written to paint a pretty picture of the world; instead, it contains the raw truth of humanity, which is sometimes an appalling picture (e.g., Judges 19).

Given the general reluctance to follow God, he was not overbearing on some issues while people took their first steps of faith. Some practices were "overlooked," at least at the time, perhaps to not overwhelm or distract people from the central message: the need and means to build a relationship with God. A clear example of this is the institution of marriage. God had in mind a man taking a woman as his wife, and the two would become "one flesh" (Genesis 2:24). For centuries, biblical figures, men faithful to God, were not addressed by God as they took multiple wives and concubines. Though strife often arose in polygamous marriages (e.g., Genesis 30; 1 Samael 1), this practice was addressed only centuries later when Jesus reminded his followers of God's original plan in marriage: one man with one woman (Matthew 19:4–5).

The Bible is also written for people, not for God or his angels. We don't see prose and poetry written for angelic beings, nor pages filled with ethereal thoughts as some might expect in a book about God. Rather, we see humans, even biblical leaders, with all their flaws, weaknesses, and ungodly behavior. There is only one exception to this, and that is the description of the life of the itinerant carpenter from the town of Nazareth in the first century. Though the Bible was written for many cultures with different needs, it makes the incredible claim that it's God-inspired:

> All Scripture is inspired by God and is useful for teaching, for reproof, for correction, and for training in righteousness, so that everyone who belongs to God may be proficient, equipped for every good work. (2 Timothy 3:16)

Keep in mind that parts of the Bible are some 3000 years old. Ancient books read differently than modern books. Ancient Near Eastern thought differs from modern Western thought. Cultural and societal importance changes and emphases evolve. Imagine if the ancient Hebrews had the opportunity to read some of our current literature. What would they think? The early text in the Bible covers what to us is a bewildering array of laws and regulations. But imagine if we were to expose this ancient civilization to modern music or a streaming series. Would they not be just as bewildered? Remember, context is everything, and the Bible was written for us, not to us (see Chapter 6). As Dennis Bratcher points out, we are not allowed to manipulate its meaning or take bits and pieces from it just to confirm some preconceptions.

> Of course, some views of Scripture contend that all of the Bible can be interchanged since it is all absolute and abstract truth written by God himself unrelated to any historical or cultural context. From that assumption, it follows that biblical verses can be added together like letters of the alphabet to make up different messages apart from what the verses themselves mean in context. However, it seems obvious to most that while the Bible is certainly true, that truth is incarnated within the historical and cultural milieu of ancient Israel and first-century Judaism. The very fact that the original languages of Scripture are Hebrew, Aramaic, and Greek ought to be a significant clue that the text is historically conditioned. That means that the text has a historical and cultural context within which it must be understood. Since the text is not abstract truth unrelated to historical context, then we must take seriously the human dimension of Scripture, including how books are written to communicate a message. This suggests that single texts must not only be understood within a historical context, but also be understood in the literary context in which they occur.[15]

It's easy to misconstrue the story of God's relationship with humankind when it is taken out of context. The Bible recounts that in the past, relatively small numbers of people sincerely sought God—and there is nothing new today! Various pagan cultures surrounded the ancient Hebrews. These cultures fabricated their own gods; again, nothing new today. The difference is that the rules of the Old Testament were more about mitigating some of the truly hideous practices of the surrounding society than about laying down instructions for a perfect society.

The Hebrew culture faced threats to their survival. If their faith was to survive and not be diluted, they had to be protected somehow. As a result, some unusual, weird, or even some morally disturbing rules were laid down, at least from our perspective in the twenty-first century. While God presented laws to the Hebrews and made promises as they kept them, this relatively small group of believers was surrounded by the "p's": the Near East was polytheistic, polygamist, and patriarchal; and prostitution was practiced everywhere. False gods reigning over agricultural production required sacrifice, including babies and young children. The gods reigning over human fertility required sexual activity as part of worship. The carnal practices of some god-worshipers included having sex with temple prostitutes. The religious practices of early civilizations were widely documented. As an example, a German psychiatrist, Iwan Bloch, pointed out the following:

> This sexual religious mysticism meets us everywhere—in the religious festivals of antiquity, the festivals of Isis in Egypt, and the festivals of imperial Rome, both alike accompanied by the wildest sexual orgies; in the festivals of Baal Peor, among the Jews, in the Venus and Adonis festivals of the Phoenicians, in Cyprus and Byblos, in the Aphrodisian, the Dionysian, and the Eleusinian festivals of the Hellenes; in the festival of Flora in Rome, in which prostitutes ran about naked; in the Roman Bacchanalia; and in the festival of the bona dea, the wild sexual license of which is only too clearly presented to our eyes in the celebrated account of Juvenal.[16]

The Bible points out that the people of God should have nothing to do with these practices (Deuteronomy 23:17). Some heroes and leaders of the faith practiced some of these traditions, including polygamy. The law of Moses did not even explicitly addressed this problem; however, these same rules did address the more overarching concern of the heart, "You shall not commit adultery" (Deuteronomy 5:18 NIV).[17]

Some eighteen centuries later, Jesus reminded people of this law and redirected them to the heart of the matter. He addressed the issue of purity at a much deeper level than merely avoiding the practices of the pagans. The thoughts of our mind were the focus of the message:

> "You have heard that it was said, 'You shall not commit adultery.' But I say to you that everyone who looks at a woman with lust has already committed adultery with her in his heart." (Matthew 5:27–28)

Likewise, divorce was tolerated for centuries but was never the plan of God, and Jesus made this clear:

> They said, "Moses allowed a man to write a certificate of dismissal and to divorce her." But Jesus said to them, "Because of your hardness of heart he wrote this commandment for you. But from the beginning of creation, 'God made them male and female.' 'For this reason a man shall leave his father and mother and be joined to his wife, and the two shall become one flesh.' So they are no longer two, but one flesh. Therefore what God has joined together, let no one separate." (Mark 10:4–9)

Though some of the laws made absolute sense from a civil society point of view (e.g., "You shall not murder" [Exodus 20:13]), others were made to address a current issue, and they may seem bizarre to us; for example: "You shall not boil a kid in its mother's milk" (Exodus 34:26b).[18] Most likely, this rule addressed a pagan practice. The intention behind the rule was that the people of God would not be tempted to follow some of the more destructive practices that were tied to it.[19]

So, the way to read the early law is to understand the context. Furthermore, the way to read any part of the Bible is to understand the context. Deciphering some of the ancient Hebrew practices and their purposes does take a bit of effort, but there are many good books available for the interested reader. I have included some examples in the bibliography.[20]

Biblical direction changes over time, though the heart of the matter remains the same. Parents who raise their kids understand this concept. Typically, it's not the parents who need to mature (although there are exceptions there!), it's the kids. So parent-child advice changes over time. If we keep this in mind, many of the seemingly strange rules and regulations of the Bible make much better sense to us.

The relationship—more than just a here-and-now experience

I remember the first time our school hosted a dance. The boys sat on one side of the hall, the girls sat on the other, the chaperones placed themselves in the corners, and we all felt awkward. The music was a sappy song from the 70s, and I remember I just wanted the ground to open up and swallow me. I think most of us felt similarly. That social experiment lasted only an hour or so, but to be honest, we all survived just fine. I think I even managed to have some conversation with a girl who was more or less normal. But this event showed me that when relationships are not well defined, or you think they are not well defined, things get awkward.

Remember, the Bible is not about science. There are instances where scientific insight seems to appear, but this is incidental and not the focus. The Bible is all about relationships, primarily between God and us; this relationship is different from ones we are familiar with in our day-to-day experience. In terms of relationships, we generally are concerned with the here and now. But a relationship with God is not just the here and now, it also looks toward its continuance after death. This long-term view means that we might expect some surprises.

A tsunami hits a coastal region and indiscriminately kills hundreds of thousands of people, both young and old.[21] The atheist sees this, and though grieved, is not perplexed by anything spiritual. The event was an act of nature and, though tragic, can be explained simply by geophysics, oceanography, and being at the wrong place at the wrong time. But a believer can be very confused by it, thinking that if God had the power to do something about the tsunami, certainly he would have, and perhaps thinking that he does not have the power.

A central theme in the Bible is imitation—us imitating Jesus. Jesus' life was riddled with many hardships (Hebrews 5:7). All his hardships, including the most dramatic one at the end of his short life, could have been avoided had God the Father so decided, but they were not. The very nature of Jesus' suffering was all in line with God's ultimate plan. What was being achieved was far more significant than merely having Jesus appear on Earth and recount wise sayings and beatitudes. And much to our inconvenience, God expects that we see suffering and hardship through the lens of Jesus Christ's life: "For to this you have been called, because Christ also suffered for you, leaving you an example, so that you should follow in his steps" (1 Peter 2:21).

As painful as this is, imitating Jesus, according to the Bible, is precisely how we will get the most out of this life (John 10:10). Hardship prepares us for a relationship with God in this life and the afterlife. Suffering is a profound concept and requires that we soberly consider what it means as we study the life of Jesus. At the same time, the Bible also teaches that God is not limited in scope. He is not only concerned with sincere, agonizing prayers but welcomes prayers for small things (Ephesians 6:18a). I can pray for patience while stuck in traffic and expect (through faith) that God will provide me with willpower and perseverance. I can even pray that the traffic lets up and makes my life easier that day. Meanwhile, God is listening to a mother who is wrestling with him in prayer over the loss of her six-month-old baby in the aftermath of the tsunami. In all cases, God is not overwhelmed; he is concerned with the relationship, here, now, and forever—with me, mother, and baby.

Though my first-world problems are trivial compared to others, God is not irritated if I ask him for relief for such a small issue as a traffic snarl as he is listening intently to the mother on the other side of the world going through agony. Understandably, many have a hard time with this concept. According to the Bible, God's mind is so superior to mine that I cannot even put myself in the right framework to understand what he is thinking. I don't claim this as an excuse, nor does it imply that Christianity has no response for such tragedies; it merely says that God's view on the world is at another level. The Bible says, "For my thoughts are not your thoughts, nor are your ways my ways, says the Lord" (Isaiah 55:8).

Of course, Isaiah did not imply that we do not mourn and weep over tragedies. Jesus also grieved when faced with friends who had died (John 11). He certainly could have prevented his friend Lazarus from dying, and he did eventually raise him from death (John 11:38–44). We sometimes lose the point of this touching story in the wake of its dramatic elements. Jesus brought back his friend Lazarus to life, and God in the flesh wept (John 11:35), identifying with our pain. But Jesus was showing a glimpse of that relationship into the afterlife.

> Then Jesus told them plainly, "Lazarus is dead. For your sake I am glad I was not there, so that you may believe. But let us go to him." (John 11:14–15)

I would assume that most of the eyewitnesses to this event did not get this point; they were simply happy to see Lazarus alive again. I suppose if I were there, I would have felt the same. But remember, the relationship God wants with us goes well beyond that of a man rising from a tomb. Lazarus lived perhaps another 20 years and then died again. God was thinking about eternity, not just the momentary happiness of friends. If we are to grasp the full meaning of the Bible, we must look beyond the here and now toward the "there and forever after."

Free will

The concept of free will is often brought up by fellow scientists, philosophers, and skeptics in our conversations about God. People often oversimplify this topic. (Admittedly, my patience has been tested in discussions of this sort—my bad!). Remember, we are not dealing with particles, planets, or even penguins, but humans, and we must appreciate free will if we are to understand who we are, especially in terms of our relationship with God.

Do you have kids? If not, maybe a nephew or niece? At the very least, you were once a kid yourself; perhaps you still are. Think about the difficulties and challenges parents have with their children. Though it's hard for us, we parents allow our children to take risks and to suffer from their mistakes or the mistakes of others. Society does not look down on parents as evil for allowing some freedom to their children. As I pushed off my daughter on her colorful little bike for the first time, the one without training wheels, I was cautious, as a concerned dad should be. I followed behind her as she pedaled, knowing that I was ready to catch her if she tipped over. I did catch her a few times as she wobbled back and forth, but eventually, she rode too fast for me, and I had to trust her balance.

My wife and I could only watch on the side as she circled around the driveway, going faster and faster as she learned her new skill. She did have a few crashes and scraped a knee, but nothing serious. We had to let our daughter ride on her own; had we not, she would have been deprived of this important life skill and missed out on fun with her friends. Freedom is difficult, but it's essential for humans.

God also gives us free will, and this is far riskier than allowing a daughter to ride a bike without training wheels. Humans can do serious damage in exercising their free will. I have watched my son play in a club basketball league. He plays very well, considerably better than I ever could. On the sidelines, I filmed his best shots on my tablet, keeping the good shots and deleting the missed one. I easily get wired up when I see an aggressive kid on the opposing team become a bit pushy. One time my son took an elbow in the nose and bled. I was angry! But he healed; his team won; dad and son both ended up happy.

But scraped knees and bloody noses cannot even compare with the atrocities of life. God allows humans to use their well-developed cerebral cortex to mastermind evil. The horror of an assailant with a modern automatic weapon marching through the hallways of an elementary school is a risk God took in allowing free will. But the ability to choose poorly is part of the human experience; without it, humans would not be humans. Furthermore, without free will, we would not have the opportunity to respond to God in a meaningful way. Unselfish love, which comes from understanding, taking risks, and self-sacrifice, is the rose petals on an otherwise thorny plant.

Yes, I suppose an all-powerful God could call for instantaneous and miraculous incarceration of the mass murderer as soon as he arms up to commit a crime. That kind of intervention from God would reduce tragedies, for sure. But the mind of the murderer would not have changed: he would still be bent on killing and perhaps even more frustrated that he did not achieve his goal. But most of us would be totally satisfied with that solution; I know I would be. But as I said above, God is not like us. He is concerned with how we think. Jesus made the interesting point:

> "You have heard that it was said to those of ancient times, 'You shall not murder'; and 'whoever murders shall

be liable to judgment.' But I say to you that if you are angry with a brother or sister, you will be liable to judgment; and if you insult a brother or sister, you will be liable to the council; and if you say, 'You fool,' you will be liable to the hell of fire." (Matthew 5:21–22)

This passage is from Jesus' iconic Sermon on the Mount, which is full of how God views things. God is just as concerned with our thoughts as with our actions. If we consider it, we would want our kids to have the right thoughts as well, not just perform the right actions because it makes our life more convenient. Of course, this example does not, in the least, diminish the agony of losing a loved one following a vicious act by someone exercising their free will; I will offer no quaint saying that makes apologies here. But I do point out that the atheist provides no solution to a tragic act except that it's statistically bound to happen.

God foresaw every single act caused by humans or nature, whether it be a scraped knee, a bloody nose, bloodshed by the dozens in the school hallways, or a tsunami that kills hundreds of thousands. We might have forgotten the execution of Jesus 20 centuries ago, but the point is that this single act allowed for reconciliation for all who accept his message, despite our poor choices in life. The painful memories left in us following the death of a loved one are certainly no less than God's painful memory of his very own Son, tortured and hung on wood, all as a result of our free will. And since God lives outside time and space, his memory of this event does not fade with time; it's a vivid reality for eternity, past, present, and future.

The light and the darkness
We know what we know about our universe because the physical laws of nature hold up in one place the same as in another. If my pencil rolls off my desk at my home in Virginia, I know that it will fall at a rate of approximately 9.8 meters per second per second. But if I go to the top of Les Diablerets (Switzerland), a mountain of 3176 meters (10,531 feet) near my relative's home, the situation is different. I can drop my pencil, and it will fall at a slightly reduced rate.[22]

The forces acting on my pencil have not changed; gravity is still gravity. But at 3176 meters I am further from the earth's center, so there is more distance between me and the mass of the planet. The local gravity has changed (because of distance) but not the fundamental gravitational force (see Chapter 4). If I went to the moon and dropped my pencil, it would still fall, but at the significantly lower rate of 1.625 m/s^2, but again, the fundamental force of gravity has not changed, only its application. If you go to any place in the observable universe, the fundamental force of gravity will be the same. Only the behavior of the pencil will vary depending on the amount of mass near it.

Why do I discuss gravity in a chapter on spiritual concepts? For the same reason that the debate about George Carlin's "invisible man in the sky" has been happening for so long. Religious practices vary with culture. Some practices follow the Bible fairly closely (e.g., the remembrance of Jesus' death); other practices vary but are, in themselves, neutral or harmless (e.g., festivals such as Christmas); still others can be outright dreadful (e.g., child sacrifice). But some people group all these practices into a big basket called "religion" overseen by the "invisible man in the sky." Notwithstanding George Carlin's view on God, I can easily identify with his dismay of religion. The story of God's relationship with man seems messy, not because God is messy, but because man is messy. Let's face it, the earth, though beautiful in all its manifold ways, is messy and complicated because of us.

The term for "sin" in the Greek is used as an archery term—to miss the mark.[23] Spiritually speaking, the bullseye mark is that of perfection. Only God can achieve perfection all the time, and of course, Jesus' life manifests that perfection. The Bible also explains that, very early on, humans (Adam and Eve) decided to use their free will to sin, and since that moment, they have been shooting the arrow not just a bit off the bullseye, but all over the place.

On the other hand, God is perfect: we say that God is light (1 John 1:5). Although God is ultimately responsible for creating everything, including light and darkness (Isaiah 45:7), we freely choose to sin and therefore enter darkness (John 3:19). It's this choice through free will that is problematic. And if we voluntarily choose to be in darkness, then God has no choice but to be separated from us, even though God is also love (1 John 4:16).

There is something about the reality and consequence of sin that is again reminiscent of science. We use evidence to test theories and ideas; measurement or observation must support what we propose. But more fundamental than the physical sciences is the language of science: mathematics.

Mathematics exists outside of any physical measurement, and it's identical in all places and at all times. An expression, such as $2 + 2 = 4$, is the same here, there, everywhere, and forever. It does not even matter if we are talking about our own universe or any of the proposed alternative universes in multiverse theory. It does not even matter if we are considering time-before-time (e.g., before the Big Bang). The equation, $2 + 2 = 4$, is always true; it will never be 5: it cannot be.

No sane scientist will argue about math (though I know a philosopher would, much to my chagrin). But spiritually, we can use the same rationale: God cannot be associated with darkness. Put another way, God cannot be associated with sin: my sin, your sin, or anybody's sin. The dark circle in the Venn diagram below, representing sin, could never overlap the clear circle, representing God. But the problem with humanity is that we all have, as adults, made deliberate decisions to sin (Romams 3:23). Therefore at least some of our Venn diagram representing our spiritual existence is in sin, in darkness. It's totally irrelevant how much we have sinned; all that matters is that we have sinned, and therefore, we cannot be associated with God, since God is only light.

This grave situation has nothing to do with God's discrimination, nor was dealing with sin an afterthought for him. God was not caught by surprise when humans sinned the first time; he foreknew all that would happen (Romams 8:29). But the reality is that God cannot associate with darkness.

The situation in the figure is somewhat reminiscent of a mathematical *axiom*, a statement considered true. I call this the light-and-darkness axiom: in no place can any darkness ever be associated with light. For God, this is a problem: every adult human has sinned. God needed a plan, and indeed, he had a plan. Remember, God would not get around the rules, just as it would make no sense for him to allow for 2+2 to equal 5.

The solution was this: God would offer himself as a trade; his clear circle could be exchanged for the dark one. The dark circle would become clear, and God's circle would become dark.[24]

Jesus was that substitute (1 Peter 2:24). God would replace the original number 2 in the equation with a larger number 3 (representing God) minus 1 (representing me and my sin). The result could still be achieved: 3 - 1 = 2, and this still gets me to 2. I could subsequently put this 2 back into the original equation, 2 + 2 = 4. Despite my contribution of the -1, all is restored. This is the way spiritual substitution works. Jesus' perfect life allowed him to come back from death (Acts 2:24), but he did go through it in our place. As previously mentioned, since God is outside time and space, his substitution, and the pain he went through in darkness and death, remains etched on him as a permanent reality (Revelation 5:6).

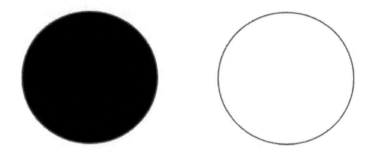

Venn diagram: Light and darkness, sin and God – part 1

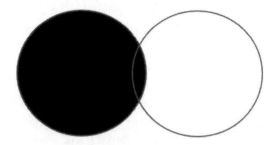

Venn diagram: Light and darkness, sin and God – part 2: this can't happen

God of the Gaps? | 207

As we leave this section, it's safe to say that the Bible was written in response to sin—had sin never existed, the Bible would be superfluous. Furthermore, without sin, there would be no misconception about God and no need to convince anyone that the heavens reveal him.

But what if?

There are times when a scientist must speculate; there is nothing wrong with that. It's troubling not being able to understand, measure, or test an idea. Though we want to understand, we will perhaps never know the answers to some of the most profound questions about our existence. So we speculate. This trait runs deep in the human psyche.

The chance that we are here "by chance" is slim, to say the least. Hopefully, the numbers expressed in Part II make us think about this more deeply. But a relatively recent idea has appeared among the scientific community that attempts to remove any of the above issues. This idea is supported by the combination of an early universe, the *inflationary theory,* and the theory of the very small, energetic state of matter of the early universe, *string theory*. Both theories are beyond the scope of this book, but we can mention that in combination, they might allow for an essentially infinite number of universes, something called the *multiverse*.[25]

The idea is that in our very early universe, space and time grew larger, very quickly, driven by a source of positive potential energy. As new space is created, small energy structures, or strings, are created. Strings are thought to be tiny waves of energy and can take on various forms, assuming different patterns. Each pattern allows for a different value of a fundamental constant (Chapter 4). Space grew so fast that "bubbles" of space separated from other parts of space. Light, traveling faster than anything else, is presumed not to have the time to go from one bubble to another. The result is many, many bubbles—perhaps an infinite number of bubbles—each one becoming a separate universe with its own sets of fundamental constants.

Theorists who like this concept believe that most of the multiverse is composed of empty, lifeless universes where the constants of nature are wrong or a planet like our planet Earth does not exist, or life never spontaneously appeared. But obviously, we are here today, and you are reading this sentence and perhaps sipping a coffee. So the multiverse concept allows, statistically speaking, at least one universe such as ours to exist as a random act of nature, and this is possible since there are, theoretically, so many universes available.

Infinity is a remarkable number (if it can even be called a number). The combination of inflation and string theories creates so many universes that, somewhere out there, all the fundamental values of physics are appropriate for life to happen, a planet such as Earth exists, and life has had the opportunity to start on its own. As a matter of fact, we would expect an identical universe out there, entirely like ours, but in that universe, you would be enjoying a black tea, not a coffee. As amazing as these scenarios may seem, this is precisely the thinking of some cosmologists. This explanation satisfies naturalists, since it offers an alternative and entirely natural explanation to the problems listed in Part II.

On the other hand, the multiverse theory is not a testable one and likely never will be, so it takes a certain degree of faith to follow the multiverse "religion." But some will remain ardently skeptical of God, and perhaps one day, a ripple in the cosmic microwave background (CMB) could hint that our universe's bubble came in a close brush to another bubble early on.

Though very skeptical, there are days when I speculate and let my imagination run wild: What if cosmologists discovered a clear and undeniable ripple in the CMB? What if astrobiologists found a viable Earth II? Suppose someone found life on another planet? Would any of these discoveries negate God? All these questions are fun to ask. Interestingly, if one took the time and stepped through the logic of each idea, we would open some fascinating possibilities for science, but not a single idea would negate God.

Even if the multiverse were proven true, it would not explain how something came from nothing. Furthermore, the question of "why" remains unanswered by the multiverse theory unless you are satisfied with "just because." Indeed, many are satisfied with "just because," but perhaps one day, curiosity on the "why" may emerge.

We will never have access to all knowledge, no matter how much we study and measure. There are places, scales of size and time, that are simply not accessible to us. In an analogy, the famed explorer Ferdinand Magellan explored a new route to the East Indies, and following his death, his expedition discovered a way to the circumnavigation of the Earth. But unlike Magellan and his expedition, no vessel could lead us to the answers to the ultimate questions of the universe. We cannot "go where no man has gone before"; what Magellan and his crew did cannot happen here. Science has always had and will always have unreachable goals. Despite scientific "leaps" in the past century, we are approaching the fundamental knowledge limit with regard to our universe's boundaries. The ensemble of naturalist theories, though well considered, will ultimately be forced to rely on faith.

Where to now?

Science will keep discovering new things, despite the diminishing returns as time goes on (remember our spring-and-mattress example in Chapter 2, note 9). But discovering and understanding God is a process that goes on forever, and, in my opinion, there are no diminishing returns. The relationship between two people, for example, has many layers, but as the Bible explains, a relationship with God provides for never-ending discovery. Each new layer with God is an opening to a new and deeper reality. Flipping the "God of the gaps" idea on its head, more discoveries of our universe reveal more, not less, about a Creator. Nothing is wrong with being a healthy skeptic; one cannot be a scientist otherwise.

We have used the term "proof" from time to time in this book. Strictly speaking, a proof of God does not exist—it never existed and never will. But the evidence for God is available today, and indeed, there is abundant evidence to support the case for a Creator. But now the reader must decide about that next step. At some point, everyone must choose their direction: people must weigh the available evidence, make tests, decide, and then base their faith either on a Creator or on a philosphy.

We are all searching for something on our paths. If you're on a hill and continue to go up, you will reach the top, no matter what direction you take. Yes, there are subpeaks and mountain valleys, but hopefully, you get the picture. The only ones who fail to reach the top are those who deliberately go downhill (those who stop asking questions) or who simply stop moving (those who think none of this is for them).[26] If you made it this far, keep reading, keep asking questions, keep seeking, and yes, keep practicing what you already understand. I assume you are on that hill and climbing—don't stop.

Chapter summary and questions

There will always be skeptics and critics of the Christian faith. True, the way Christianity has been practiced over the centuries has invited significant cynicism; however, if one is interested in peeling away the layers of religious confusion, the truth of Christ can be discovered, as well as the necessity of the cross.

Questions to ponder:
- Can you think of other gods of the twenty-first century that we did not mention in this chapter?
- Can you think of a hypocritical trait of which you have been guilty? By chance, is this a trait that irritates you when you see it in others?
- Do you consider yourself a serious student of your religion? If you are a Christian, how much time do you spend in your daily effort to understand God better? If you are not a Christian, how much time do you spend in the pursuit of your religion? Is religious confusion diminishing in your life, increasing, or just staying the same? Does this need to change? How can you change it?
- Have you ever thought: "If God could just show me this one miracle, I would believe"? Has God answered your request? What was the effect on your faith?
- If you could play God for a day, how would you handle giving free will to humans knowing that they may abuse this gift? (Remember: A genuine relationship requires free will.)

NOTES
[1] Hedone: the Greek goddess of pleasure, enjoyment, and delight.
[2] Haphaestus: the Greek god of blacksmiths, metalworking, carpenters, craftsmen, artisans, sculptors, metallurgy, fire, and volcanoes.
[3] Athena: an ancient Greek goddess associated with, among other things, wisdom.
[4] Zeus: the Greek god of the sky and thunder, who rules as king of the gods.
[5] Agnostos Theos: the god of the unknown and from which the term "agnostic" is derived.

[6] Victor J. Stenger, "Is the Universe Fine-Tuned for Us?" PDF, University of Colorado, archived from the original PDF on 2012-07-16.
[7] Henry Drummond, *The Ascent of Man* (New York: J. Pott, 1904), 333.
[8] The BioLogos Foundation, "Are gaps in scientific knowledge evidence for God?" BioLogos, originally retrieved January 6, 2015, revisited in Wiki August 2018.
[9] Dietrich Bonhoeffer, "Letter to Eberhard Bethge, 29 May 1944," *Letters and Papers from Prison*, ed. Eberhard Bethge, trans. Reginald H. Fuller (New York: Touchstone, 1997), 310–312, translation of *Widerstand und Ergebung* (resistance and surrender) (Munich: Christian Kaiser Verlag, 1970).
[10] As mentioned earlier, here is where the author departs from the intelligent design views.
[11] Jim Boucher, "Can God Create A Rock So Heavy That He Cannot Lift It?" (2013) https://thereforegodexists.com/can-god-create-a-rock-so-heavy-that-he-cannot-lift-it/, last accessed August, 2018.
[12] C.S. Lewis, *The Problem of Pain* (originally published in 1940), OCR by Van Wingerden (in the public domain, 2016), 12.
[13] Einstein postulated that the speed of light c was constant (300,000 km/s or 186,000 miles/second).
[14] C.S. Lewis, *Mere Christianity*, Book IV, Chapter 2, "The Three-Personal God" (New York: Macmillan, 1952).
[15] Dennis Bratcher, "Word of Faith and 'Commanding': Contextual Analysis of Matthew 21:21," The Synoptic Problem: The Literary Relationship of Matthew, Mark, and Luke, http://www.crivoice.org/commanding.html accessed August 03, 2018.
[16] Iwan Bloch, *The Sexual Life of Our Time in Its Relations to Modern Civilization*, translated from the sixth German edition by M. Eden Paul, MD (London: Rebman, 1909), 106.
[17] Also, it's helpful to remember context here: the Hebrews were in contact with God as he spoke through Moses on Mt. Sinai.
[18] To be clear, a goat's kid, not your kids. The pagans were bad, but not all the time.
[19] There are a few possibilities for this command. A good discussion is found here: Daniel Edgecombe, "Cooking a young goat in its mother's milk," (May 2, 2018) http://living-faith.org.
[20] Especially the bibliography for Chapter 6.
[21] I was in Thailand only a month after the great Indian Ocean tsunami of 2004. Panic and shock could still be felt in the country's main airport following this devastation—an event that saw waves up to 100 feet (30 m) and was responsible for killing nearly 230,000 people.
[22] Gravity is slightly less at the top of the mountain. So if air resistance can be ignored, objects would fall more slowly at the mountain summit.
[23] The term *hamartia* derives from the Greek *hamartánein* (ἁμαρτάνειν), which means "to miss the mark" or "to err." According to J.H Thayer, this term has been associated with sin "by omission or commission in thought and feeling or in

speech and actions." J.H. Thayer, *Greek-English Lexicon of the New Testament* (New York: Harper, 1887), online at Google Books.

[24] Needless to say, God did not become completely sin. The concept of the trinity, the three persons of the godhead, Father, Son, and Spirit, is a profound biblical truth (Matthew 28:19; 2 Corinthians 3:17; 1 John 5:7–8; John 14:9–11, 10:30–36) but beyond our current study.

[25] Lisa Grossman, "Multiverse Gets Real with Glimpse of Big Bang Ripples," *NewScientist* (March 18, 2014), https://www.newscientist.com/article/dn25249-multiverse-gets-real-with-glimpse-of-big-bang-ripples/ accessed October 22, 2018.

[26] This most profound concept (and the last word for this book) is from my brilliant and lovely wife, Valérie.

Epilogue – My Own Journey

Seven and a half years after Einstein had left us, I took my first breath. My upbringing, though very different from the great scientist's, had a few similarities with his. I grew up with Christian instruction. When I reached my teen years, I was able to work through some of the same misconceptions about God that Einstein faced. My spiritual path was nonetheless circuitous and even challenging at times.

I had a passion for learning, and through frequent use of the school library and public libraries, I had access to popular science books and read them with enthusiasm. It quickly became clear to me that my pursuit would center on astronomy and physics. I was probably somewhat more religious than Albert Einstein, at least after I passed my twelfth year, but this was more out of duty than curiosity. For the most part, my religious beliefs coexisted with my scientific knowledge, though I was curious and asked many questions about the "possibilities of the impossible" we find in the Bible.

I have heard it said that the Christian life is a marathon, and I agree. But the process of becoming a Christian can also be an exploit of endurance. My spiritual journey is somewhat akin to that of a scientist making a serendipitous discovery while looking for something else. I was not necessarily looking for God (the God defined in the Bible), but I was nonetheless curious about him. There were times when I made good choices and made natural progress toward God; at other times, I made excuses, acted selfishly, or made poor choices; the latter made the entire journey more complicated, though evidently, not impossible.

I came to know God in four steps. Again, while proceeding through these steps, I had no idea that I was on some spiritual peregrination. I merely took in what I could understand at each step and asked questions all along the way. My early spiritual strides were mostly quiet and methodical and accompanied by what I would call "chats with God."

The first step came in my youth. I was exposed to the Bible, not so much by my own effort, but by just following my family. My parents brought me to a local church, and I learned about the basic structure of the Bible and the general characteristics of God. This instruction was undoubtedly helpful and perhaps not too dissimilar to what Albert Einstein was exposed to before he reached age 12.

Learning about God in this step was a bit like learning science in grade school: teachers gave some of the facts, but did not emphasize relevant equations. I learned the basics: God was God and the Almighty God could easily do things beyond our understanding. On the other hand, instruction did not stress significant biblical concepts such as discipleship, repentance, sin, and biblical salvation. None of this partial education bothered me or even seemed to hinder me at the time.

The second step came when I found the need to pursue God on my own. At the age of 17, I started reading the Bible. I began at the very beginning, Genesis 1. Each day I read a chapter or so and continued until I had read the entire book. When I got to the end, Revelation 22, I flipped back to Genesis and began the process again. I did this three times, continuing through college. Overall, this step took about eight years.

Unlike Einstein, I did not acquire a distaste for the Bible as a teen; the "naïve" and "childlike" stories that Einstein rejected did not bother me, though it was clear to me that these stories could not be reconciled with science. On the other hand, what helped me was the understanding that the Bible was not a coverup story; not all the accounts were rosy. Ironically, as I read about atrocities committed by many leaders in the Bible and of the horrible consequences of sin, I gained respect for the book. Though it laid out the truth as absolutes, there seemed to be nothing artificial or plastic-religious about it.

By pursuing God on my own, I inadvertently prepared myself to quickly evaluate and sidestep most of the more unusual variations on Jesus' teaching. As a teenager living in the United States, I quickly learned that there were thousands of representations of his teachings, and along with those interpretations came variations on what it means to be a Christian. I learned only much later that sometimes those variations were innocuous cultural shades, but others were distracting, nonbiblical, and some even detrimental to one's faith. By reading the Bible on my own, I was able to avoid most of this confusion and was unwittingly preparing myself for life challenges that were to come.

But there was a downside to my one-person religious experience: it was easy for me to be self-deceived. Again, I was slowly realizing how difficult it was to follow Jesus on your own. Even after I had read the Bible three times, the basic concepts of Christianity, including discipleship, repentance, sin, and salvation, still did not register with me, though none of these concepts were that complicated to understand. Later, I realized that I was not paying attention as I read, or I subconsciously minimized the importance of these practical parts of the Bible and focused perhaps too much on the academic questions.

On the other hand, I was observant of the Christian community around me and their practices. Occasionally, I slipped into the back pew of one of the nearby churches in Cambridge, Massachusetts, where I lived. There were many congregations from which to choose, and I made regular visits to different ones. These visits had a mixed impact on me. I was happy to be in church, but it was hard for me to identify with a stranger who shared the pew with me. I had no idea how to (nor the desire to) exchange views about my day-to-day practice of Christianity, nor did I have the feeling that the person sitting next to me wanted to chat. The only time I interacted with someone else was when the preacher asked us to say hello to our neighbor. I said, "Hello," and that was that. I remained detached, trying to figure out what people were thinking, but making little effort to talk with others. Once I made it back to my dorm, I had "God chats" with my physics friends and, in particular, Serge; this I found decidedly more entertaining.

I did pay close attention to what the preacher was preaching in church. Some preachers gave messages that were almost identical to the one from the previous week, and this eventually wore me down. In other churches, the preacher raised his voice too much, so I did not return a second time.

During this step, I slowly lost interest in my spiritual journey, but my passion for science grew. I wanted to explore fundamental questions in astrophysics, so I targeted a career as an astronaut-scientist. I had the good fortune of pursuing my studies at the Massachusetts Institute of Technology. I worked on my first thesis at the gravity wave laboratory alongside brilliant physicists; one would eventually be awarded the Nobel Prize in Physics for his contributions to confirming this manifestation of general relativity.[1]

During my studies, I remember tirelessly going over the work of Albert Einstein. On one Friday evening, I sat in the spacious MIT

Hayden Library in my favorite quiet corner. There was a beautiful, expansive floor-to-ceiling window overlooking the sparkling Boston skyline and the Charles River. Taking a rest from homework, I read about Einstein's personal habits. As he came close to framing his general theory of relativity in 1915, he became so singularly focused that he took little time to do anything else. His clothes were wrinkled, and his crazy hair took on a shape of its own. Lost in his mental exercises, he would have the first course to his meal but then find no time to change cutlery. He used the same spoon as he went from his bowl of soup to his main dish. This was odd, I thought, but understandable. Einstein sat for hours through the day motionless, no writing, no talking, just thinking, and merely sorting the concepts out in his mind. Something about his quirky behavior and his remarkable focus sank deep into my psyche. His idiosyncratic habits inadvertently helped me understand something—singular focus. I knew I needed this quality in my academic pursuit; later I would learn that it was also critical in pursuing my relationship with God. During this time, I was not completely lost in science; I continued with my daily habit of reading a chapter in the Bible.

The third step in my spiritual journey was painful and came midway through my graduate studies. Up to that time, I faced a constant and challenging set of opportunities that propelled me forward toward my personal goals. Nevertheless, for me to embrace God, I had to go through a difficult time physically, emotionally, and professionally. Though it may seem a small handicap to many, I lost my ability to distinguish colors; therefore, I also lost one of my dreams—to pursue a career as an astronaut. Soon after this first defeat, I faced a second one. Though I was fortunate to be able to continue my graduate studies at MIT, my dream PhD research project came to an abrupt halt. I was exploring "negative energy" as a replacement in Einstein's field equations with the hope of "breaking the light barrier."[2] My PhD advisor was skeptical of this work, and I was asked to abandon it. I was traumatized. But instead of slowing down in my research, I became more determined. I decided to redouble my focus on science in my life, to regain something I felt I had lost. I spent the next few months so fixated on physics that I abandoned most of the friendships around me. Not surprisingly, a girlfriend of several years left me. All these events came together and pushed me to the brink of desperation.

It was a Thanksgiving weekend, and I sat alone in a large computer hall at MIT. I plugged away with astronomy data on a new project, but in my mind, I could not forget the pain of realizing that my advisor had rejected my thesis idea. I stared at the screen, but I no longer cared about what I was doing. I sat motionless, with my head in my hands and coffee cup next to me, waiting for my computer to reboot after an unexpected crash. The sound of the mainframe cooling equipment permeated the computer room as I tapped my foot on the raised floorboards and watched the tapes spinning in their cabinet. I became very frustrated with my decision to work on this holiday weekend instead of doing what everyone else did, go home and spend time with their families.

I made so little progress that morning that I just got up and left the lab. I went for a long walk. I crossed the Harvard Bridge and went through a nearly empty downtown Boston. I was greeted by loose newspaper and plastic trash, windswept down the trafficless streets. I continued to the south end of the city, an area known many years ago as the Boston Neck. At the top of a bridge, I looked out and then down. What raced through my mind in those few minutes was disturbing. Little did I know in my depressed state, still with lots of head knowledge but with little purpose, that I was getting much closer to understanding God. I faced a decision I had never before encountered. I needed to find sense in my life. Fortunately, instead of doing something crazy, like throwing myself off the bridge, I decided to walk back home. In that long walk, I had time to consider the few things that mattered to me, but this was extremely hard since that which mattered to me had been taken away.

The fourth step took a friend, or in my case, two friends, Ted and Pete. They were both physics graduate students who had recently started their own work toward their PhD. A week earlier, I had moved into my new office just next to theirs.

As I sat in my office that Monday morning, preparing the course work I would deliver to undergraduates later that afternoon, Ted and Pete came over to say hi. We quickly became good friends. Both of them played rugby, and all three of us lifted weights. Ted was from Hawaii; he wore Hawaiian shirts and had a social life. His girlfriend,

Yuri, used to bake cookies and bring them to the office; there were always a few for me. Pete was from Australia. He had a pleasing accent and a special gift for starting a conversation. The three of us got along very well.

A week or so after my dreadful Thanksgiving weekend and our first encounter, they invited me to an informal Bible discussion on campus. I was apprehensive at first, thinking back to the lackluster visits to the local churches. Those visits had left me with a spiritual bad taste in my mouth. But since I had very little going on in my life at that point, I decided to attend the discussion. The group leader, Mark, was another graduate student. The topic was not too heavy, and the company, mostly MIT undergrads, was very amicable. A few more weeks went by, and Mark invited me to a personal Bible study. Now I was a bit more apprehensive. The explanation that helped me was that this study could better assist me in putting the Bible into practice.

I thought to myself, this explanation did make sense, so I swallowed my pride and agreed. I studied the Bible and made some very tough decisions to reprioritize my life. Just over three months later, I was baptized. Even after four steps, 26 years, two months, and nine days, the journey had just begun. Life as a Christian can be like a marathon—but by faith, there is a golden ribbon at the finish line.

NOTES

[1] Professor Rainer Weiss, cowinner of the 2017 Nobel Prize in Physics.

[2] Several concepts would allow an object to go beyond the speed of light. One idea includes creating wormholes, tunnels through space and time. This concept has been vigorously pursued by scientists at the theoretical level, especially since the mid-1980s.

Appendix 1 – Faith: Biblical and Scientific – Two Sides of the Same Coin?

Stating the obvious, this book is about faith. But the term "faith" can be confusing; it has different meanings in different settings. For our purposes, we start with the dictionary definition of faith:[1]

> Faith, *noun*.
> 1. Confidence or trust in a person or thing
> 2. Belief that is not based on proof
> 3. Belief in God or in the doctrines or teachings of religion
> 4. Belief in anything, as a code of ethics, standards of merit, etc.
> 5. A system of religious belief
> 6. The obligation of loyalty or fidelity to a person, promise, engagement, etc.

Your definition of faith will vary, but the religious world tends to use definitions 3 or 5, the secular world sometimes uses definitions 4 and 6, and religious skeptics use definition 2. Definition 1 is perhaps the most useful in this book.

By viewing faith according to definition 1, "biblical faith" and "scientific faith" have more in common than one might think. Biblical faith trusts in someone (God) and scientific faith trusts in something (experience or observations). Biblical faith is confirmed when the source of that faith reveals itself, or rather himself; scientific faith is confirmed when new observations confirm that which was once assumed. These manifestations of faith are really two sides of the same coin, they both have a source, and they both need confirmation; absent of these, they would both be blind. Faith on both sides of the coin is characterized by moments of waiting, revelation, and in some cases, simple trust in the source while waiting for the next confirmation. Some aspects of faith, both biblical and scientific, can never be confirmed, at least in our finite existence or on this side of death (e.g., What happened before the beginning of time? What are the details of the afterlife?)

Between the two, biblical faith is more easily identified and framed. We find the classic source for this faith in the eleventh chapter of the book of Hebrews, and it sounds somewhat like the first definition from the dictionary: "Now faith is the assurance of things hoped for, the conviction of things not seen" (Hebrews 11:1).

Christians view this passage as the powerful central text of faith, though admittedly, without further context, it's easy to confuse this definition of faith with definition 2, belief in something without proof. To clarify this, it's helpful to continue reading.

> Indeed, by faith our ancestors received approval. By faith we understand that the worlds were prepared by the word of God, so that what is seen was made from things that are not visible. (Hebrews 11:2–3)

The chapter goes on for 40 verses, and though we do not copy them here, they are worth reading. Hebrews 11 gives many examples of people who exercised their faith, and we commend them for doing so; sometimes responses to faith were given right away (verses 17, 23, 28, 29, 35a), but at other times not (verses 7, 11, 13, 22, 35b–39). These examples show their faith woven into the trust they had with their God. The people mentioned here had incredible faith—Abel (verse 4), Enoch (verse 5), Noah (verse 7), Abraham (verse 8), together with Isaac and Jacob (verse 9), and Sarah (verse 11). These people were commended for their faith, along with many others (verses 22–38). Furthermore, their faith was strengthened as they took steps based on what they already knew (e.g., Gideon—Hebrews 11:32 and Judges 6:14); their faith was not blind. Ultimately, people of faith are asked to hold on to their faith beyond this life to a goal on the other side of death. Hebrews 11 ends this way:

> Yet all these, though they were commended for their faith, did not receive what was promised, since God had provided something better so that they would not, apart from us, be made perfect. (Hebrews 11:39–40)

That is faith in terms of the Bible, but what about science? To be certain, you will not find a single chapter in any book paralleling Hebrews 11 for science, but that does not mean science is absent of faith. Over the centuries, science has updated itself with new tests, new observations, and new measurements, so that which was once "faith" in this or that theory has been updated as fact as new evidence came in. I suspect that most scientists would be hesitant to label their trust in a particular theory as faith. To be fair, a scientist only trusts in a current working theory until someone performs a new observation. For the most part, religious reverence is absent here.

On the other hand, if you were to wander through the basements of a large research library, you would find thousands of scientific articles bound and stored. These articles represent the recorded path of scientific progress. We expect a steady flow of new knowledge to appear and add to the knowledge that already exists. But as we pointed out in Chapter 2, this can go on only so long. Eventually, our limit to observations will constrain fundamental science. One day, science will no longer be able to make significant new measurements to push the boundaries of fundamental knowledge any further. Science will rely totally on past experiences and theories. Eventually, those theories will become a philosophy—naturalism is an outcome of this progression. To this, I say, is there any difference in which side of the coin we are considering? Faith is on both sides; only the object of that faith differs.

NOTES
[1] http://www.Dictionary.com *faith*

Appendix 2 – "In the Beginning"

For those who are scientifically minded and have been in Bible culture for many years, it's understandable that we expect to find extraordinary scientific facts in the Scriptures. For example, if God oversaw the writing of the Bible (2 Timothy 3:16–17), he would have introduced concepts that should amaze us and likewise confirm that it's God-inspired. After all, wouldn't God want us to be impressed?

To be sure, yes, God wants to impress us. He wants us to be confident that our faith is real and based on solid and substantial evidence. But biblical evidence provided by history and prophecy notwithstanding, we might want to see more from the Bible than what is offered. This is especially true for the twenty-first-century (scientific) Bible believer.

When the modern reader starts reading Genesis, we ask the natural questions: When did it all begin? And how did it all begin?[1] We head to the very first verse of the Bible. It appears that God created something from nothing. The NIV reads: "In the beginning God created the heavens and the earth" (Genesis 1:1).

In these first words of the Bible, it appears that God creates something from nothing; this is the doctrine of *creatio ex nihilo* (creation from nothing) and is the prima facia implication of God's creation scheme. This concept appears to be supported by the book of Hebrews.

> By faith we understand that the worlds were prepared by the word of God, so that what is seen was made from things that are not visible. (Hebrews 11:3)

We saw in Chapter 7 that the worldview of the ancient Hebrews is very different from ours. Creation from nothing did not necessarily imply something from nothing, but rather from unorganized chaos to an organized universe. Rabbinic scholars concur that in the ancient worldview, chaotic waters, not absolute vacuum, filled space: "[The] idea that the sky-dome was made of congealed water makes eminent sense in terms of creation out of watery chaos."[2]

Other scholars have pointed out that the Genesis writer was also thinking about going from past to future, not of going from the

Appendix | 223

invisible to the visible and certainly not from absolute vacuum to space, time, and matter.[3] It's conceivable that the concept of *creatio ex nihilo* was too sophisticated for ancient civilizations, and therefore Genesis 1:1 would seem to convey that God works with the formless void (already present at the beginning).[4]

Had we not considered these arguments, the timeline for Genesis 1:1 may look like the top graph in the figure below. This timeline represents the Big Bang model with creation starting at t = 0, the *de facto* beginning of time—this is probably not what the writer had in mind. The bottom figure shows a timeline that is more representative of the text. But we still need to be cautious: the very concepts of graphs, timelines, and plotting are modern ones stemming from Western thinking, so we should probably not read too much into a quantitative discussion of the beginning.[5]

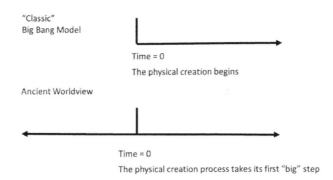

Two concepts of time for our universe

Sequencing is another issue that arises in the scientific discussion of biblical creation. If one reads the Genesis account, there is a series of events that broadly aligns with modern theory. But it's helpful in this discussion to start with the Genesis account:[6]

Day one
> In the beginning when God created the heavens and the earth, the earth was a formless void and darkness covered

the face of the deep, while a wind from God swept over the face of the waters. Then God said, "Let there be light"; and there was light. And God saw that the light was good; and God separated the light from the darkness. God called the light Day, and the darkness he called Night. And there was evening and there was morning, the first day. (Genesis 1:1–5)

Day two
And God said, "Let there be a dome in the midst of the waters, and let it separate the waters from the waters." So God made the dome and separated the waters that were under the dome from the waters that were above the dome. And it was so. God called the dome Sky. And there was evening and there was morning, the second day. (Genesis 1:6–8)

Day three
And God said, "Let the waters under the sky be gathered together into one place, and let the dry land appear." And it was so. God called the dry land Earth, and the waters that were gathered together he called Seas. And God saw that it was good. Then God said, "Let the earth put forth vegetation: plants yielding seed, and fruit trees of every kind on earth that bear fruit with the seed in it." And it was so. The earth brought forth vegetation: plants yielding seed of every kind, and trees of every kind bearing fruit with the seed in it. And God saw that it was good. And there was evening and there was morning, the third day. (Genesis 1:9–13)

Day four
And God said, "Let there be lights in the dome of the sky to separate the day from the night; and let them be for signs and for seasons and for days and years, and let them be lights in the dome of the sky to give light upon the earth." And it was so. God made the two great lights—the greater light to rule the day and the lesser light to rule the night—and the stars. God set them in the dome of the sky to give light upon the earth, to rule over the day and over the night, and to separate the light from the darkness. And God saw

that it was good. And there was evening and there was morning, the fourth day. (Genesis 1:14–19)

Day five

And God said, "Let the waters bring forth swarms of living creatures, and let birds fly above the earth across the dome of the sky." So God created the great sea monsters and every living creature that moves, of every kind, with which the waters swarm, and every winged bird of every kind. And God saw that it was good. God blessed them, saying, "Be fruitful and multiply and fill the waters in the seas, and let birds multiply on the earth." And there was evening and there was morning, the fifth day. (Genesis 1:20–23)

Day six – part 1

And God said, "Let the earth bring forth living creatures of every kind: cattle and creeping things and wild animals of the earth of every kind." And it was so. God made the wild animals of the earth of every kind, and the cattle of every kind, and everything that creeps upon the ground of every kind. And God saw that it was good.

Then God said, "Let us make humankind in our image, according to our likeness; and let them have dominion over the fish of the sea, and over the birds of the air, and over the cattle, and over all the wild animals of the earth, and over every creeping thing that creeps upon the earth." (Genesis 1:24–26)

Day six – part 2

So God created humankind in his image, in the image of God he created them; male and female he created them. God blessed them, and God said to them, "Be fruitful and multiply, and fill the earth and subdue it; and have dominion over the fish of the sea and over the birds of the air and over every living thing that moves upon the earth." God said, "See, I have given you every plant yielding seed that is upon the face of all the earth, and every tree with seed in its fruit; you shall have them for food. And to every beast of the earth, and to every bird of the air, and to everything that creeps on the earth, everything that has the breath of life, I have given every green plant for food." And it was so. God saw everything that he had made, and indeed, it was very good. And there was evening and there was morning, the sixth day. (Genesis 1:27–31)

Day seven
Thus the heavens and the earth were finished, and all their multitude. And on the seventh day God finished the work that he had done, and he rested on the seventh day from all the work that he had done. So God blessed the seventh day and hallowed it, because on it God rested from all the work that he had done in creation. (Genesis 2:1–3)

This creation account takes place in six "days," with the seventh day representing the day of rest (God is no longer creating). We assume these are allegorical days,[7] representing periods of time.[8] Do we see modern science discussed here?

Some of the events are in the correct order. For example, Genesis 1 explains the creation of matter before life, followed by the creation of lower life, and then humans.[9] On the other hand, there are also problematic issues in Genesis 1 that do not link to our modern understanding. We have already named the major obstacles in Chapter 6—it was written with the backdrop of an ancient worldview; artifacts of this worldview are present in Genesis 1, including 1) a solid dome (firmament) above the Earth, supporting an ocean (watery expanse)—nothing in our modern understanding of cosmology resembles this concept;[10] 2) light from the Sun is misordered, coming only after the creation of life;[11] 3) a misordering of animal creation: fish appeared some 400 million years before fruit trees,[12] and birds did not precede land animals.[13]

Additional sequencing problems arise if we look to the second creation account in Genesis 2. This second creation account begins in the following way:

In the day that the Lord God made the earth and the heavens, when no plant of the field was yet in the earth and no herb of the field had yet sprung up—for the Lord God had not caused it to rain upon the earth, and there was no one to till the ground; but a stream would rise from the earth, and water the whole face of the ground—then the Lord God formed man from the dust of the ground, and breathed into his nostrils the breath of life; and the man became a living being. (Genesis 2:4b–7)

The day of creation in this account appears to merge days from the first account.[14] In summary, one must be careful when attempting to superimpose our twenty-first-century science on the Genesis creation account.

NOTES

[1] These questions are different from the concerns of the Israelites as Genesis was being penned. See Copan and Jacoby, *Origins*, 26.

[2] Gier, "The Three-Story Universe."

[3] This is the point of G. W. Buchanan, "The author's concern for the unseen was not primarily that which was invisible or intangible, but that which was future, that which had not yet happened." George W. Buchanan, *To the Hebrews (The Anchor Bible)* Vol 36 (New York: Doubleday, 1972), 184.

[4] W. R. Lane, "The Initiation of Creation," *Vetus Testamentum* Vol. 13, Fasc. 1 (Jan 1963): 63–73. Also noted by Gier, the terms "void" and "nothing" sometimes refer to watery chaos, paralleling Genesis 1:1 (see Jeremiah 4:23; Isaiah 40:17, 23). Gier, "The Three-Story Universe."

[5] The bottom graph does not imply that the Hebrew writers were considering an infinite universe, nor a cyclic universe, nor any physical explanation, before the "beginning"—they simply were not thinking in these terms. Hence, the Hebrew writer neither supports the Big Bang model nor refutes it. But for the curious, there are some physicists, for example, Rodger Penrose, who support a cyclic universe. A popular article on this can be found here: Cosmology Research Update, "New evidence for cyclic universe claimed by Roger Penrose and colleagues," 21 Aug 2018, https://physicsworld.com/a/new-evidence-for-cyclic-universe-claimed-by-roger-penrose-and-colleagues/

[6] There are actually several creation accounts in the Bible: 1) Genesis 1:1–2:3, 2) Genesis 2:4–26, 3) Psalm 89:9–12, 4) Proverbs 8:22–31, 5) Isaiah 40:21–28, 6) Ezekiel 28:12–15, 7) John 1:1–3, and 8) Colossian 1:15–17. See Copan and Jacoby, *Origins*, 90–91.

[7] A literal interpretation (without consideration of context) forces a 6 × 24-hour creation week; this is young earth creationism. There are two challenges in using this interpretation: 1) it forces the text to mean something that it was never intended to mean, and 2) a literal six-day creation account describes a universe that is only a few thousand years old—an interpretation completely contrary to scientific evidence; the universe is billions of years old.

[8] Much could be said on the notion of "day" and the Bible. We refer the reader to two sources: 1) Copan and Jacoby, in *Origins*, 69, point out that the literal view

creates nonsensical situations on creation day six, especially in terms of the first day of life for Adam; 2) John Oakes, in *Is There A God?*, 116, stresses the fact that *yom* is translated 1611 times in the Old Testament into day (KJV) and many of the translations connote a nonliteral meaning, or something other than a strict 24-hour clock time. Some other notions of time include "today," used 30 times or "age," used six times or even "perpetually," used twice.

[9] Beyond these general points: the universe did have a beginning (the singularity of the Big Bang), light (pure energy) was the only component of the universe at the beginning, the Earth was created before animals, and animals were created before humans.

[10] Many (including the author) have invested significant amounts of time and energy attempting to show that the firmament is an atmosphere. The Hebrew term *raqia* is derived from the root *raqqə ʻ* (רָקַע), meaning "to beat or spread out thinly," e.g., the process of making a dish by hammering thin a lump of metal, "Lexicon Results Strong's H7549 – raqiya`," Blue Letter Bible, retrieved 2009-12-04.

[11] This problem has been argued in the following way: perhaps light from the Sun only appeared at the surface of the Earth on day three, though the Sun was present on day one. The delay came because of a thick haze that covered the Earth in its early stages until about 2.5 billion years ago with the production of oxygen in photosynthesis. On the contrary, light of a longer wavelength still would have reached the Earth's surface. The problem with these types of speculations is that there is no evidence in the text that this was happening; we are only superimposing our modern science.

[12] Some have speculated that this problem goes away by generalizing "seed" in Genesis 1:12. The term seed comes from the Hebrew term *zera* (זֶרַע) and might refer to the more general term "offspring." Using this definition, the problem that fruit-bearing trees being out of order with animal creation in day five disappears. But again, there is no evidence in the text supporting this interpretation.

[13] Modern science says the sequence should be: 1) fishes, 2) reptiles (including ground reptiles), 3) birds, and 4) mammals, while the Genesis sequence is: 1) fishes and birds (day 5), 2) ground reptiles and mammals. Again, some have speculated that land animals refers to only domestic animals, appearing around 9000 years ago. Yet again, there is no evidence to support this interpretation.

[14] More on the second creation account can be found in Copan and Jacoby, *Origins* (2018), 68–69.

Appendix 3 – Scientific Concordism and Skating

Biblical inerrancy is fundamental to the Christian faith. Paul writes in his letter to Timothy,

> "All scripture is inspired by God and is useful for teaching, for reproof, for correction, and for training in righteousness so that everyone who belongs to God may be proficient, equipped for every good work." (2 Timothy 3:16–17)

Also, Peter writes,

> No prophecy ever came by human will, but men and women moved by the Holy Spirit spoke from God. (2 Peter 1:21)

These scriptures indicate that God inspires the Bible. Does this mean that the Bible should always be taken literally? Some conservative Christian circles believe this to be true, but literal interpretation everywhere is problematic.

Jesus used hyperbole. As he rebuked the Pharisees for their hypocrisy, he said,

> "You blind guides! You strain out a gnat but swallow a camel!" (Matthew 23:24)

Jesus also uses hyperbole when challenging us not to sin,

> "If your right eye causes you to sin, tear it out and throw it away; it is better for you to lose one of your members than for your whole body to be thrown into hell." (Matthew 5:29)

I would argue that no one takes the Bible literally, and if anyone did, we might expect camel consumption and Christians wearing eye patches. But hyperbole is not the only problem; the Bible is also 30 percent poetry. If we want to accept all Bible poetry as literal, then we will be faced with non-sensical concepts (e.g., God having wings and feathers [Psalm 91:4]).

On the other hand, Christians should take scriptures seriously; this implies that we carefully consider what is literal and what is imagery. We must also consider what is essential and what is background information.[1] We adapt the following figure from Denis Lameroux's book, *Evolutionary Creationism*.[2] The Bible contains statements on theology, history, and science, and each corresponds to different realities. The center of the figure represents an overlap of the subject areas and is a subject of interest. In the reference mentioned above, that theme of interest was the origins of humans; however, "X" could represent any topic we might find in the Bible, including farming, fishing, engineering, accounting, management, or finance. We can extract biblical examples for almost any of these topics and find supporting statements of theology, history, and science in the Bible.

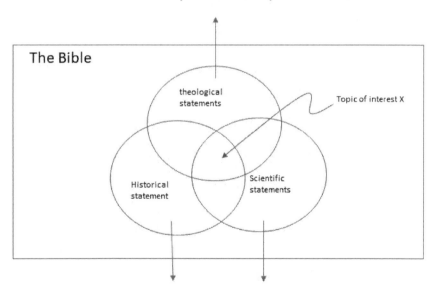

General view of biblical concordism (Source: Adapted from D. Lamoureux, Categories of Biblical Concordism, Evolutionary Creationism)

Of course, we assume that the theological statements made in the Bible are central in importance, and therefore the concordance between the theological statements and spiritual reality is strong. Historical statements have been supported in this work and described in Chapter 8 and Appendices 4 and 5. However, we should not take all history of the Bible literally, as pointed out in the ancient cosmogony in Genesis. What about science? We have already discussed science in the context of Genesis and considered the prescience cosmogony in the interpretation of Genesis 1. More generally, scientific concordism attempts to harmonize various science topics and statements made in the Bible. Some have identified these statements as a biblical science "proof" since these texts supposedly make statements about the earth that predate any science that could explain the phenomenon. As a result, some cite these verses as miraculous indications where the Bible sheds light on twentieth-century science. We find one of the more popular verses in the book of Job.

> He spreads out the northern skies over empty space; he suspends the earth over nothing. (Job 26:7, NIV)

This passage seems to suggest that our planet is hovering in space; an implication is that gravity resulting from the earth-sun interaction and the centrifugal force resulting from the earth's movement around the sun effectively "suspend" the earth in the void. The challenge with this view, as discussed in Chapter 7, is that the writer of Job has in mind the ancient Genesis cosmogony. The writer may be describing the earth this way as it rests in the ocean of the three-tiered universe. This description explains the writer's perspective, but what about our perspective?

Denis Lamoureux points out that these verses are "ripped out of Scripture and their ancient scientific context, and then [are conflated] with modern scientific ideas."[3] If indeed this verse points to a scientific breakthrough, then we would expect these verses to be supported by other scriptures. In terms of the earth floating in space, this is not the case. For example, in the book of Job, we have the image of the earth upheld by pillars:

> He shakes the earth from its place and makes its pillars tremble. (Job 9:6, NIV)

As pointed out in Chapter 7, this verse, along with others (Job, 38:4–6; 1 Samuel 2:8; Psalm 75:3) allude to earth's foundations and its pillars, both artifacts of ancient cosmogony. The contemporary Bible reader may assert that these verses are metaphorical, while Job 26:7 is literal, but this is a forced interpretation. An analogy may be helpful here.

One of the more confusing notions in freshman physics is the idea of rotational inertia. Instructors often demonstrate the concept using familiar objects, including figure skaters. In their best sketch, they would draw out a skater with their arms extended and then pulled in, increasing their rate of spin, as shown in the figure.

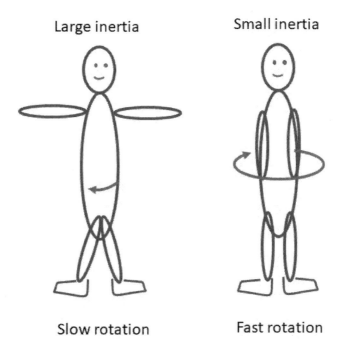

Rotational inertia demonstrated by a skater

Physics textbooks commonly use sketches to convey concepts; skaters demonstrate complex ideas (rotational inertia). It would make no sense for the textbook to use a more abstract illustration (particle fields) to convey the same concept. A complex illustration would only confuse the reader (freshman engineering student) who is desperately trying to understand the material (rotational inertia) to pass the class.

Now suppose a curious engineer and skating enthusiast opened the physics textbook and came across this picture, noticing some details about the skater. The engineer might understand that the textbook was illustrating the fine points of figure skating. It's possible that the engineering and skating enthusiast may be more interested in the illustration about skating than the physics behind the picture.

Now we rephrase the above: Earth, sky, water, and ancient cosmogony are used to demonstrate a complex idea (God's nature). It would make no sense for the Bible to use a more complex idea (modern cosmology) to convey the same concept; this would only confuse the readers (ancient Hebrews) who were trying to understand the material (God's nature). But is it possible that the physics textbook writer might have slipped the skating figure in on purpose for other reasons? Perhaps he was thinking about the skating enthusiast. Whatever the case, details of the figure skater, would be no more relevant to the engineering student than is the biblical description of the earth for the modern Bible reader.

One can review other verses in the Old Testament in the same way that we discussed Job 26:7. These concepts include the water cycle (Job 36:27; Amos 9:6b; Jeremiah 10:13; Psalm 135:7; Ecclesiastes 1:7), deep-sea springs (Job 38:16), atmospheric circulation (Ecclesiastes 1:6), and the unfathomable number of stars in the sky (Genesis 22:17; Jeremiah 33:22).

NOTES

[1] Douglas Jacoby has used the phrase, "at face value," to incorporate all types of biblical genre, literal, and nonliteral, in such a way that all Bible passages are taken seriously, but in their proper context, poetry included.
https://www.douglasjacoby.com/q-a-1242-why-don-t-you-take-the-bible-literally/

[2] Dennis Lamoureux, *Evolutionary Creation: A Christian Approach to Evolution* (Eugene, OR: Wipf & Stock, 2008).

[33] Denis Lamoureux, "Ancient Science in the Bible," *I Love Jesus & I Accept Evolution* (Eugene, OR: Wifp & Stock, 2009).

Appendix 4 – A Proof of Chronology: The Dead Sea Scrolls

Something remarkable happened in 1946. Muhammad edh-Dhib, his cousin Jam'a Muhammed, and a friend, Khalil Musa, were out in the wilderness of the Wadi Qumran on the northwest corner of the Dead Sea in present-day Palestine. One of their goats had wandered off, perhaps into one of the many caves on the hillside. As they searched for it, they did not realize what was about to happen— arguably one of the most significant archaeological finds of all time. As shepherds, one of their main tasks was to keep their sheep and goats from straying. A flick of rock would guide the animals from going where they should not go.[1]

Late one day as the flocks climbed high on the rocky slopes:

> One of the young shepherds tossed a rock into an opening on the side of a cliff and was surprised to hear a shattering sound. He and his companions later entered the cave and found a collection of large clay jars, seven of which contained leather and papyrus scrolls.[2]

At first, the shepherds did not know what to do with the parchments. They attempted to sell them to a dealer in Bethlehem named Ibrahim 'Ijha. It was reported that Ibrahim dismissed the scrolls as worthless, thinking that they probably were stolen from a synagogue. Eventually, the scrolls were sold to a Syrian Christian for seven Jordanian pounds, the equivalent of about $325 (2020 dollar value). The scrolls then passed hands several times, and the news began to spread that they were an extraordinary discovery.[3]

In the succeeding year, a total of seven manuscripts were discovered. These ancient documents were housed in jars in a cave at the Qumran site, what is now famously called Cave 1. Centuries ago, parchments were stored in earthenware jars to protect them. The dry climate preserves these documents so well that many were completely legible two thousand years later.

In the following months, the archaeologist John C. Trever took notice and interviewed the men and discovered more scrolls in subsequent expeditions. Several years of productive archaeology followed these initial discoveries.

Scholars examining the Dead Sea Scroll fragments at the Rockefeller Museum, formerly the Palestine Archaeological Museum (Source: Abraham Meir Habermann, Public Domain)

Trever identified some of the first scrolls to be from the writings of the prophet Isaiah. Between 1946 and 1956, archaeologists discovered many more scrolls and fragments.[4] Eventually, they uncovered 981 manuscripts, most written on parchments and some on papyrus. Besides documents, other articles were found, including cisterns, pottery, and debris from a dining facility. The caves also revealed a scriptorium and a cemetery.

Over time, Trever and others understood that they were uncovering the home of the Jewish sect, the Essenes. The Jewish historian, Flavius Josephus, documented this sect; apparently, it had large numbers in Roman Judaea, but inferior to the population of Pharisees and Sadducees. The Essenes were "dedicated to voluntary poverty, daily immersion, and asceticism (their priestly class practiced celibacy)."[5] Most scholars agree that the Essenes did not write the Dead Sea Scrolls, but were perhaps more concerned with housing the documents and living a monastic existence. Remarkably, their communities preserved not one, but multiple copies of the Hebrew Bible. Through their efforts, they preserved a significant portion of the Old Testament, as well as many ancillary writings. The documents they maintained were dated as early as 300 BCE. After news of the discovery got out, Bedouin treasure hunters and archaeologists unearthed hundreds of additional scroll fragments from 10 nearby caves.

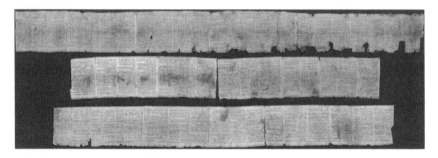

The Great Isaiah Scroll found at Qumran (Source: The Israel Museum, Jerusalem, public domain.)

The treasure of the Dead Sea Scrolls has several layers, including one of material riches. The Copper Scroll is one of the more intriguing. Hebrew and Greek letters are chiseled onto metal sheets and describe 64 underground treasure locations around Israel. Some have speculated that the treasures came from the Second Temple (70 CE), and the local Essenes hid the valuables before its destruction. No one has ever recovered these treasures; many believe the Romans pillaged these in the first century CE.[6]

Unfound stores of silver and gold notwithstanding, the real treasure of the Dead Sea Scrolls is the scientific evidence of their meaning—this cannot be underestimated. The single Isaiah Scroll, for example, is 1000 years older than the oldest authoritative Hebrew and

Aramaic text of the Tanakh (canonical collection of Jewish texts, which is also the textual source for the Christian Old Testament). Known as the Great Isaiah Scroll or 1QIsa, it contains all 66 chapters of the original text and is housed in the Shrine of the Book Museum in Jerusalem (see figure above). But far beyond the Isaiah Scroll, the ensemble archaeological find included fragments from nearly every book of the Old Testament.[7] Other documents discovered include sectarian regulations, the Community Rule, and other religious writings that also do not appear in the Old Testament.[8]

The dating of the scrolls is genuinely remarkable: no scholar has dated this scroll to later than 100 BCE. Four carbon-14 studies give calibrated dates ranging between 335 and 107 BCE. There are only minor variations from the Masoretic text in it. According to Miller Burrow:

> Of the 166 words in Isaiah 53, there are only 17 letters in question. Ten of these letters are simply a matter of spelling, which does not affect the sense. Four more letters are minor stylistic changes, such as conjunctions. The three remaining letters comprise the word "LIGHT," which is added in verse 11 and which does not affect the meaning greatly.[9]

Furthermore, this text is also supported by the Septuagint (LXX).[10] In brief, the Dead Sea Scrolls essentially prove that the book of Isaiah is accurate and that we can have confidence in the version of the Old Testament in our possession today.[11]

We find significant Old Testament prophecies of Jesus in the book of Isaiah, particularly in chapter 53, which contains the eerie foretelling of the demise of Jesus. This twentieth-century discovery gives us documentation that contains a description of Jesus written more than seven centuries before his life and copied between one and a half to three centuries before he walked the Earth. One passage in Isaiah describes Jesus' punishment on the cross, and incidentally the hope for humanity:

> Surely he has borne our infirmities
>   and carried our diseases;
> yet we accounted him stricken,

> struck down by God, and afflicted.
> But he was wounded for our transgressions,
>   crushed for our iniquities;
> upon him was the punishment that made us whole,
>   and by his bruises we are healed. (Isaiah 53:4–5)

This text provides evidence that the description of Jesus' death was "miraculously" foretold centuries before the event.

NOTES

[1] Harry F. Thomas, "How the Dead Sea Scrolls Were Found," *Biblical Archaeology Review* 1:4 (December 1975).

[2] J. Cohen, "6 Things You May Not Know About the Dead Sea Scrolls" (August 12 2018), https://www.history.com/author/jennie-cohen, accessed August 2018.

[3] J.C. Trever, *The Dead Sea Scrolls* (Piscataway, NJ: Gorgias, 2003).

[4] "Hebrew University Archaeologists Find 12th Dead Sea Scrolls Cave," The Hebrew University of Jerusalem, February 8, 2017 https://new.huji.ac.il/en, accessed August 2018.

[5] F.F. Bruce, *Second Thoughts on the Dead Sea Scrolls* (London: Paternoster, 1956).

[6] Ibid, 74 and Cohen, "6 Things You May Not Know About the Dead Sea Scrolls."

[7] Only the Book of Esther was not recovered, and it has been argued that this was perhaps because the Essenes did not celebrate Purim, the holiday festival based on the Book of Esther.

[8] Bruce, *Second Thoughts on the Dead Sea Scrolls*, 74, and Cohen, "6 Things You May Not Know About the Dead Sea Scrolls."

[9] Millar Burrows, *The Dead Sea Scrolls* (Chicago: Moody 1986), 304.

[10] The Septuagint LXX (70 in Roman numerals) is informally called the Greek Old Testament and is the earliest extant Koine Greek translation of the Hebrew Scriptures. The Septuagint was translated between the third and second centuries BCE.

[11] Geisler and Nix, *A General Introduction to the Bible,* 263.

Appendix 5 – Messianic Prophecies

We have seen in Chapter 8 that the building blocks for validating the Bible have come from evidence that can be tested and measured with historical information. This next set of evidence enters the realm of the supernatural; there is simply no way to get around this point. At this juncture in our investigation, we cannot avoid exploring the supernatural aspect of the Bible and God, and this is evident in prophecies, where historical evidence meets faith head-on. So we shall examine these prophecies, comparing the prophets' writings and their fulfillment. There are many examples of prophecy in the Bible, but we focus on the most important, the prophecy of the incarnation of God, the Messiah, Jesus Christ; his life; and his death.[1]

Messianic prophecy itself is less astonishing than the sheer number of prophecies. Predictions of the future are the mainstay of many religious movements, including those that fall into "Christendom." The tests for prophecy are simple: Do the predictions come true, and are those prophecies specific enough to be credible? Failed attempts to forecast the future fall into one of two categories: either predictions are simply wrong a statistically significant fraction of the time, or their forecasts are so vague that they are useless.

- Isaiah 7:14 – He would be born of a virgin (Lk 1)
- Micah 5:2 – He would be born in Bethlehem (Lk 2:4)
- Genesis 49:10 – He would be born of the tribe of Judah (Mt 2)
- Psalm 78:2 – He would speak in parables (Mt 13:34)
- Zechariah 9:9 – He would ride on the colt of a donkey (Mt 21)
- Isaiah 61 – He would heal the brokenhearted (Lk 4:18)
- Isaiah 53:3 – He would be rejected by his own (Jn 1:11)
- Isaiah 53:7 – He would stand silent before his accusers (Mk 15:5)
- Psalm 22:18 – Some would cast lots for his robe (Jn 19:23–24)
- Psalm 22 – His hands and feet would be pierced (Mk 15:25)
- Psalm 22:1 – He would experience a separation from God (Mt 27:46)
- Zechariah 11:12 – He would be sold by enemies for 30 pieces of silver (Mt 26:15)
- Isaiah 53:9 – He would be buried with the rich (Mt 27)

The Bible contains hundreds of predictions, many of them specific. Except for prophecies that still postdate us (e.g., Jesus' return to earth), all have been realized. Over 300 direct prophecies were made about Jesus' life, and they were made approximately 400 to 1000 years before the events happened; again, all have come to pass. A few examples of recorded prophecies and fulfillment are given here:

Jesus died by crucifixion, a method of unusual cruelty and brutality. Victims were nailed to a cross and left to die. They hung on the wood structure for hours or even days before succumbing to death. The book of Psalms, written around 1000 BCE, prophesied Jesus' death. In Psalm 22, specific details explain his death (piercing his hands and feet; his body bared to onlookers; his clothes divided):[2]

> Dogs surround me,
>   a pack of villains encircles me;
>   they pierce my hands and my feet.
> All my bones are on display;
>   people stare and gloat over me.
> They divide my clothes among them
>   and cast lots for my garment. (Psalm 22:16–18 NIV)[3]

The Persians conceived crucifixion in 300–400 BCE, and the Romans further developed it into a punishment for the most serious of criminals. Hence, when David was writing Psalm 22, this form of capital punishment would not exist for another 700 years!

If the prophecies mentioned above are sound, they must also pass the test of authenticity: Was the subject of prediction aware of the prophecy, and could he have influenced the outcome? The famous words written by David read, "My God, my God, why have you forsaken me?" (Psalm 22:1) Jesus uttered these same words: "About three o'clock Jesus cried out with a loud voice, 'Eli, Eli, lema sabachthani?' that is, 'My God, my God, why have you forsaken me?'" (Matthew 27:46; Mark 15:34).

One might legitimately argue that Jesus had these lines rehearsed so that he knowingly would help fulfill one of the 300 prophecies about his life. But this seems like a lackluster way to validate an Old Testament prophecy, knowing that his life would end in a moment. If all other prophecies were not accurate, all would end for him, invalidating him as the Messiah. On the other hand, in the case of dozens of other prophecies, someone couldn't possibly manipulate the results. These prophecies include Jesus' birth in Bethlehem, prophesied in Micah 5:2; his coming from the line of Judah, prophesied in Genesis 49:10; and perhaps the most bizarre prophecy, his being born of a virgin, prophesied in Isaiah 7:14.[4] Some scholars have estimated that Jesus Christ fulfilled nearly 300 references to 61 specific prophecies of the Messiah. The chances that one person could fulfill these many predictions by chance is utterly remote.[5]

Evidence of the resurrection

According to the Bible, Jesus' life was premised on his being a sacrifice, a substitute who would take the punishment for people's sin. God could have chosen many other ways for Jesus to die and still equated his death with the required sacrifice. But as it was, Jesus found his demise in crucifixion, a Roman form of torture that ultimately led to his death on the cross as recorded in the four Gospels: Matthew 27, Mark 15, Luke 23, and John 19.

Had Jesus been killed and not resurrected, the entire premise of faith in God would be worthless (1 Corinthians 15:17). Our faith would be worth no more than the stone altars and statues noticed by Paul in his missionary journey through Athens (Acts 17:23). On the other hand, if Jesus did resurrect from the dead, and evidence showed this, not only does this fulfill prophecy, but it gives anyone with an earnest desire to take a step of faith that very opportunity.

As difficult as it is to imagine the agony of death on the cross, the good news of the resurrection of Jesus that followed is the central hope of Christianity. There is no doubt that the resurrection of Jesus was a miracle, and the pivotal one on which all Christianity hangs. So important is this prophecy that one should take time to test the evidence for what happened that Sunday morning, the third day after the crucifixion.

We mentioned earlier that critics of Christianity don't deny that Jesus was crucified under Pontius Pilate. Many skeptics have attempted to propose alternative explanations of the resurrection, and several Christian apologists have explored these possibilities. Some alternative possibilities include: 1) Jesus was not actually dead; 2) the women visiting him could not find his body because they went to the wrong tomb; 3) Jesus' body was stolen; and 4) eyewitnesses hallucinated Jesus' resurrection out of wishful thinking.[6] It's helpful for one not familiar with the three-day event to mentally walk through and explore these alternative explanations. The Apostle John, an eyewitness to the death and resurrection, and Luke, Greek physician and historian, both give accurate details of this event. Readers have the essential material before them to piece together the story as investigators.

On Friday afternoon, around 3 pm, Jesus was dead. To make doubly sure he was deceased, a guard plunged a spear into his side. This violent act caused a sudden flow of fluid (John 19:34). The *pericardium*, or the sac surrounding the heart and containing *pericardial effusion* (fluid), was pierced. The Roman guards often broke the legs of their victims in a procedure called *crurifragium* that would lead to rapid death. This violent act was not done to Jesus, as he was determined to be already dead, fulfilling prophecy (recorded in John 19:32–33, 36, and prophesied in Psalm 34:20). Jesus was brought down from the cross and wrapped in material for his burial (Luke 23:56). His body was covered with aromatic spices and wrapped in preserving material, estimated to weigh as much as about 100 pounds (45 kilograms).[7]

Next, a faithful disciple, Joseph of Arimathea, buried Jesus in a new tomb (Luke 23:50–53). It's important to note that the women followers of Jesus accompanied Joseph to the tomb and saw how Jesus' body was laid (Luke 23:55). Roman guards were assigned to watch the tomb and set an official seal over the stone closure—removing the official seal was punishable by death. The disappearance of Jesus' body that Sunday morning does not lend itself to other possibilities. Jesus was undoubtedly not faking his death, and the women did not visit a wrong tomb that just happened to be empty. Furthermore, the Roman guards would risk execution if they had failed their mission, letting someone steal the body. Beyond this, the Roman and Jewish leaders would have done everything to account for the body of Jesus; their motivation was to impede the new sect called Christianity.

The Romans posted a guard, who then placed a massive stone (estimated to weigh up to two tons) at the entrance to the tomb.[8] The disciples were in no shape to take on the guards and move the stone in an attempt to make it look like Jesus resurrected. They were either discouraged like Peter (Mark 14:72; Luke 22:62) or fled in fear (Mark 14:50).

When Peter and John finally made it to the tomb on Sunday morning, they noticed something strange. They saw the cloth that had been around Jesus' head and the linen wrappings (John 20:5-7), but they saw no corpse. What readers often forget in the resurrection story is that the apostles did not understand the resurrection until later (John 20:9). This means that the disciples were not motivated to steal the body, even if they possibly could. Matthew records the guards' report of this incident (Matthew 28:11–15). Their story corroborates the facts that they did not know what happened to the body and accepted a bribe:

> While they were going, some of the guard went into the city and told the chief priests everything that had happened. After the priests had assembled with the elders, they devised a plan to give a large sum of money to the soldiers, telling them, "You must say, 'His disciples came by night and stole him away while we were asleep.' If this comes to the governor's ears, we will satisfy him and keep you out of trouble." So they took the money and did as they were directed. And this story is still told among the Jews to this day. (Matthew 28:11–15)

This story seals the concept that both the guards and the chief priests were privy to the resurrection event, and they intentionally chose to create an alternate motive. This account also clearly states that people were widely circulating the false narrative. Some people would have been able to compare the report perpetrated by the chief priests to another account that would soon be available to them.

There were many witnesses to the resurrection. After appearing to Mary (John 20:11–18) and two disciples at Emmaus (Luke 24:13–31), Jesus appeared to all the disciples apart from Thomas (John 20:19–25). A few days later, Jesus also appeared to Thomas (John 20:26–29). After this, he appeared to the fishermen at the shore (John 21:1–14) and then to 500 people at once (1 Corinthians 15:3–8). Then Jesus appeared to James, then to all the apostles, and finally, to Paul (1 Corinthians 15:7–8). It's difficult to attribute all these sightings of Jesus to individual hallucinations.

Before becoming one himself, Paul was not a friend of Christians, yet he also encountered Jesus, an event that led to his conversation. Furthermore, Paul says that among the five hundred witnesses, most were still alive at the time of his writing (1 Corinthians 15:6). Therefore, a reader of Paul's letter would have had the opportunity to query an eyewitness of the resurrection. If some of those readers had heard the conflicting message created by the chief priests, they would have eyewitnesses to consult to challenge that false narrative.

Finally, the most significant piece of evidence of the resurrection, the one that could only come by an individual convinced of the resurrection, is the power of their changed life. I have learned about and even witnessed scientific discoveries in my scientific career. Like others, I got caught up in the excitement of making small discoveries in the fields of astrophysics and nuclear physics. But discovering the nexus of evidence leading up to the resurrection cannot be overvalued.

The disciples clearly understood the significance of this discovery, and perhaps the best way we know they were not merely fabricating the resurrection is that their lives were utterly transformed. They went from the simple lives of fishermen, to depressed or cowardly men immediately following Jesus' death, to brave men. Most, if not all, of Jesus' apostles, apart from John, went to an early death.[9] Substantial evidence supports that Peter, Paul, James the brother of Jesus, and James the son of Zebedee died as martyrs. The Roman authorities killed believers in excruciating ways. Some were crucified, beheaded, pierced by spears, stoned to death, or burned to death.[10] Had any one of these men simply said they did not see Jesus alive, they may well have lived to see another day. Yes, some followers did eventually deny they were followers of Jesus and were able to save their own lives, but the Christian community never altered the question of the resurrection. Many Christian families were tortured for acknowledging their faith. Mothers saw their young die in the Roman arena, and spouses saw their loved ones killed.

Those convinced of the resurrection refused to stand down or deny what they knew. The evidence of the resurrection convinced the Christians, allowing them to take a step of faith the world had never before seen.

NOTES

[1] I fully realize that some reading this will find the concept of God becoming flesh and then allowing himself to be executed not to make any sense, and as a matter of fact, the concept represents a complete paradox. Nonetheless, I encourage you to keep reading. Remember, this concept was also a stumbling block for the first-century Greeks and Jews (1 Corinthians 1:22–24).

[2] The verb "pierced" in Psalm 22:16 clearly points to crucifixion. The words "pierced" and "lion" are remarkably similar in Hebrew. A minority of the Masoretic texts (tenth through eleventh centuries) contain the term "lion." However, in the Dead Sea Scrolls, which predate most other Hebrew texts by over a thousand years, the term is unmistakably "pierced." In addition, the oldest Syriac, Vulgate, Ethiopic, and Arabic versions also use the term "pierced." The same is true in the Septuagint, which was completed approximately 200 years before the birth of Christ. E. Würthwein, *The Text of the Old Testament: An Introduction to the Biblia Hebraica* (Eerdmans, 1995).

[3] We point out that despite the remarkable parallels with Jesus' death, we cannot be certain whether this psalm was prophetic or just referring to David's own misery (metaphorically speaking), or both. This passage was written by King David (circa 1000 BCE). A helpful discussion by John Oakes, "Is David talking about himself in Psalm 22:16–18? How do you know this is a messianic prophecy?" can be found at http://evidenceforchristianity.org/is-david-talking-about-himself-in-psalm-2216-18-how-do-you-know-this-is-a-messianic-prophecy/ (October 17, 2011), last accessed October 4, 2019. Oakes maintains, "The inference of the evidence is that David is not talking about himself in this famous passage."

[4] A study of messianic prophecies can be found here: Douglas Jacoby, "Messianic Prophecy," July 31, 2013, https://www.douglasjacoby.com/messianic-prophecy/, last accessed June 14, 2019.

[5] For simplification, suppose that each prophecy about the Messiah were given a chance of ½ (that is, there is a statistical 50-50 chance that the prophecy is correct). Though this value may be very generous, it's useful in this exercise. Taken together, say on 61 independent prophecies, the combined probability that the prophecies are correct by chance is: $(½)^{61}$ or 1 divided by 2,305,843,009,213,693,952, or essentially zero.

[6] There are various lists that cover alternative theories for the death and resurrection of Jesus. This particular list is inspired by the one generated by Josh McDowell in "Evidence for the Resurrection" an extract from *Evidence that Demands a Verdict*, Craig Blomberg, *Jesus and the Gospels* (Nashville, TN: B&H, 2009), and Douglas Jacoby, *Compelling Evidence for God and the Bible: Finding Truth in an Age of Doubt* (Eugene, OR: Harvest House, 2010).

[7] McDowell, *Evidence that Demands a Verdict*.

[8] Ibid.

[9] John was exiled on the Island of Patmos, perhaps in the 90s CE. Also, according to Tertullian (c.155–240 CE) in his *On the Prescription of Heretics*, John might have even been boiled in oil, but did not die.

[10] Curtis, Ken. "Whatever Happened to the Twelve Apostles?" https://www.christianity.com/church/church-history/timeline/1-300/whatever-happened-to-the-twelve-apostles-11629558.html (April 10, 2010), accessed June 2019.

Acknowledgements

Over the years, many people, some living, some passed, have played a role in my life and subsequently helped this work come together. In my early years, the works of Albert Einstein and Carl Sagan inspired me the most. I was introduced to them in my high school physics class, through the episodes of *Cosmos*. Their work, writings, and views of the world struck a chord in my mind. I am sure Carl Sagan did not intend this as an outcome, but his ethereal monologue came across to me in an almost spiritual way. Oddly, he helped me to think about God. But the real thanks go to Mr. Butterworth, my high school physics teacher extraordinaire. His attention to us, getting us to ask questions about the world around us, was what I most remember about him. But he was not the only physics teacher who helped me. A few years later, I was also blessed to work with Professor Rainer Weiss, my first research advisor. He was kind, encouraging, and dedicated to discovering that which seemed impossible to discover, gravity waves. By observing him, I learned what it takes to persevere in a seemingly impossible effort. And it was also during this time in my life that I was wrestling with the topics in this book.

I am incredibly indebted to other scientists and academics who are living examples in their faith. Drs. Douglas Jacoby and John Oakes both come to mind. They have offered their time and knowledge as input for this book, and both are incredible examples, professionally, academically, and spiritually—thank you, gentlemen! Also, thank you, John Clayton, a fellow scientist and Christian. I admire your work. You have inspired me through your presentations to pursue this line of apologetics as a way to give back to the community—thank you again, Mr. Clayton! Others have encouraged me to pursue this book and share these ideas in my own neighborhood of Northern Virginia—thank you, Randy McKean. Neighbors and friends include Lael Kaplan and Grant Schneemann, who have listened to my thoughts and ideas over the past five years as I wrote this book—thank you for your patient listening and helpful discussions.

This book would not have been possible without the help of many readers and editors. I heartily thank Jade Olsen for her wise input and helpful suggestions. I thank Amy Morgan for her diligence and keen eye in helping shape the manuscript. I would like to thank Mark and Heather Halverstadt for their input to the book; an additional thanks to Heather for her helpful tips in making this book more accessible to the Christian community and to nonscientists. Brett Kreider, I very much appreciate your helpful and insightful suggestions. I thank Nicole Cosey for her help in formatting the book.

I also thank Bill and Betty Zachary, readers of this book, editors, and, most of all, loving parents. Speaking of my family, my siblings, Dawn, Kurt, and Mark, have together, along with my parents, allowed me to be a free thinker. They created a safe atmosphere as I grew up and fostered a place where I could ask questions and share my thoughts without being judged. I am forever thankful for you all. I am grateful for my dad's advice to read the first few pages in Genesis in their proper context. This was a time (in the 1970s) when church and the science lab were two very separate and detached places (indeed they are still separate and becoming more and more distant!).

Finally, I want to thank my immediate family. My children, Calypso and Ulysse, now teenagers, hung on with me as we talked about topics in this book—thank you for your patience and perseverance. Finally, I am forever indebted to my wife, Valérie. She encouraged me when I got stuck on ideas and trapped in my own thinking. She reminded me why I am writing this book—to help people read and understand another book, one far superior to this one. Thank you, Valérie, for your love and perseverance and for reminding me about that which is most important.

Bibliography

Chapter 1
Darwin, Charles. *The Origin of Species by Means of Natural Selection, or the Preservation of Favoured Races in the Struggle for Life (6th ed.)*. London: John Murray, 1972.
Dictionary of Christianity and Science, edited by Paul Copan, Tremper Longman III, Christopher L. Reese, and Michael G. Strauss. Grand Rapids, Michigan: Zondervan, 2017.
Gould, Stephen J. *The Panda's Thumb*. New York: W.W. Norton & CO., 1982.
Lamoureux, Dennis. *Evolutionary Creation: A Christian Approach to Evolution*. Eugene, Oregon: Wipf & Stock, 2008.
_____. *I Love Jesus & I Accept Evolution*. Eugene, Oregon: Wifp & Stock, 2009.
_____. *Evolution: Scripture and Nature Say Yes*. Grand Rapids, Michigan: Zondervan, 2016.
Orgel, Leslie. *The Origins of Life*. 1973.

Chapter 2
Ackerman, James S. *Distance Points: Essays in Theory and Renaissance Art and Architecture*. Cambridge, Massachusetts: MIT Press, 1991. ISBN 978-0262011228.
Doudna, G. "Carbon-14 Dating." In *Encyclopedia of the Dead Sea Scrolls* Vol.1. Edited by Schiffman, Lawrence, Emanuel Tov, and James VanderKam. New York: Oxford University Press, 2000.
Drake, Stillman. *Galileo At Work*. Chicago: University of Chicago Press, 1978. ISBN 0-226-16226-5.
O'Connor J.J., and E. F. Robertson, eds. *Abu Ali al-Hasan ibn al-Haytham*. MacTutor History of Mathematics archive. Scotland: School of Mathematics and Statistics, University of St Andrews, 1999. Last accessed 2008-09-20.
Page, Don. "How to Get a Googolplex." 3 June 2001. Archived 6 November 2006 at the Wayback Machine.
Selin, Helaine, ed. "M," in *Encyclopaedia of the History of Science, Technology, and Medicine in Non-Western Cultures 1*, 1667. New York: Springer, 2008. ISBN 9781402045592.

Sharratt, Michael. *Galileo: Decisive Innovator.* Cambridge, Massachusetts: Cambridge University Press, 1994. ISBN 0-521-56671-1.

VanderKam, James C., and Peter Flint. *The Meaning of the Dead Sea Scrolls.* New York: HarperOne, 2002.

Chapter 3

Bacon, Francis. *De Interpretatione Naturae Prooemium.* Complied and published in 1653.

Baierlein, Ralph. *Newton to Einstein,* 201–202. Cambridge, Massachusetts: Cambridge University Press, 1992.

Dobrzycki, Jerzy, and Leszek Hajdukiewicz. "Kopernik, Mikołaj" in *Polski słownik biograficzny* ("Nicolaus Copernicus" in Polish biographical dictionary), Vol. XIV. Wrocław: Polish Academy of Sciences, 1969. The Polish Biographical Dictionary is translated: English–language dictionary of Polish biography, authored by Stanley S. Sokol and published by Bolchazy-Carducci in 1992.

Gregory, Peter L. *Laddie's Grave.* www.xulonpress, 2010.

Holton, G. J., and Yehuda Elkana. *Albert Einstein: Historical and Cultural Perspectives.* Mineola, New York: Dover Publications, 1997.

Newton, Isaac. *The Principia: Mathematical Principles of Natural Philosophy,* 1687.

Pedersen, Olaf. "The Decline and Fall of the Theorica Planetarium: Renaissance Astronomy and the Art of Printing." In *Science and History: Studies in Honor of Edward Rowen.* Wroclow: Polish Academy of Science Press, 1978. Studia Copernicana 16.

Tiner, John Hudson. *Johannes Kepler: Giant of Faith and Science.* Fenton, Michigan: Mott Media, 1977.

Chapter 4

Appenzeller, I., G Börner, M. Harwit, R. Kippenhahn, P.A. Strittmatter, and V. Trimble, eds. *Astrophysics Library* (3rd ed.). New York: Springer, 1998.

Atkins, P.W. *The Creation.* New York: W.H. Freeman, 1981.

Barrow J. D., and J. Silk. *Scientific American* 242 No. 4, 1980.

Barrow, J D., and F.J. Tipler. *The Anthropic Cosmological Principle*. Oxford: Oxford University Press, 1986.

Barrow, J. D., and D.J. Shaw. *General Relativity and Gravitation* 43. New York: Springer Science + Business Media, 2011.

Bridge, Mark, dir. *How the Universe Works*. Season 3, episode 2, "The End of the Universe." Aired July 16, 2014 on Discovery Channel.

Carr, B.J., and M.J. Rees. *Nature* 278, 1979.

Carter, B. *Atomic Masses and Fundamental Constants: 5*. Edited by J.H. Sanders and A.H. Wapstra. New York: Springer,1976.

Clayton, D., W.A. Fowler, T. Hull, and B. Zimmerman. "Neutron capture chains in heavy element synthesis." *Annals of Physics* 12 (1961): 331–408.

Clayton, Donald D. *Principles of Stellar Evolution and Nucleosynthesis*. New York: McGraw-Hill, 1968.

CODATA Value: Newtonian constant of gravitation. *The NIST Reference on Constants, Units, and Uncertainty*. US National Institute of Standards and Technology, 2015. Last accessed September 25, 2015.

Couchman D. "The Strong Nuclear Force as an example of fine-tuning for life." http://www.focus.org.uk/strongforce.php, 2010. Last accessed December 13, 2017

Davis, P. *Cosmic Jackpot: Why Our Universe Is Just Right for Life*. New York: Orion Publications, 2007.

Denton, Michael. *Nature's Destiny*. New York: The Free Press, 1998.

Dyson F. *Scientific American* 225 (1971): 52–54.

Gribbin, J., and M. Rees. *Cosmic Coincidences: Dark Matter, Mankind, and Anthropic Cosmology*. ReAnimus, 1989.

Hawking S. *A Brief History of Time*. New York: Bantam Books, 1988.

Horgan, J. *The end of Science*. New York: Perseus Books, 2015.

Hoyle, F. "The Universe: Past and Present Reflections." *Engineering and Science* (November 1981): 8–12.

———. "On Nuclear Reactions Occurring in Very Hot Stars." *Astrophysics. J. Suppl. 1* (1954): 121.

———. "Evolution from Space." Omni Lecture. Royal Institution, London, 12 January, 1982; *Evolution from Space: A Theory of*

Cosmic Creationism (New York: Simon & Schuster, 1984), 27–28.

_____, and C. Wickramasingh. *Evolution from Space: A Theory of Cosmic Creationism*, 27–28. New York: Simon & Schuster, 1984.

Lemley, B. "Why Is There Life? Because, says Britain's Astronomer Royal, you happen to be in the right universe." *Discover Magazine* (November 1, 2000).

Leslie, J. *Universes*. London: Routledge, 1989.

Lincoln, D. "Einstein's True Biggest Blunder." Op-Ed. Fermi National Accelerator Laboratory, November 6, 2015. Retrieved from http://space.com, December 8, 2017.

Livio, M., D. Hollowell, A. Weiss, A., and J.W. Truran. "The anthropic significance of the existence of an excited state of 12C." *Nature* 340:6231 (27 July 1989).

Penrose R. *The Emperor's New Mind: Concerning Computers, Minds, and the Laws of Physics*. Oxford: Oxford University Press, 1989.

Rees, Martin. *Just Six Numbers: The Deep Forces That Shape The Universe*. First American edition. New York: Basic Books, 2001.

Rozental I.L. *Structure of the Universe and Fundamental Constants*. SPI Moscow, 1981.

Salpeter E.E. "Nuclear Reactions in Stars." *Physical Review* 107, 1957.

Schroeder G. "The fine-tuning of the universe: Quotes from Steven Weinberg." http://geraldschroeder.com/. Accessed June 2018 and December 2017.

Science Buddies. "Attraction with Static Electricity: An electrically charged challenge from Science Buddies." *Scientific American*. https://www.scientificamerican.com, January 12, 2012.

Silk, Joseph. *On the Shores of the Unknown: A Short History of the Universe*. Cambridge, Massachusetts: Cambridge University Press, 2005.

Stenger, Victor J. "Is The Universe Fine-Tuned For Us?" University of Colorado. Archived from the original PDF, July 16, 2012.

Uzan, Jean-Philippe. "The fundamental constants and their variation: observational and theoretical status." *Reviews of Modern Physics* 75:2, April 2003.
Weinberg, S. *Facing up: Science and Its Cultural Adversaries*. Cambridge, Massachusetts: Harvard University Press, 2003.

Chapter 5

Barnes, Rory, and René Heller. "Habitable Planets Around White and Brown Dwarfs: The Perils of a Cooling Primary." *Astrobiology* 13:3 (March 2013): 279–291.
Barrow, John, and Frank Tipler. *The Cosmological Anthropic Principle*. Oxford: Oxford University Press, 1988.
Boehle, A. et al. "An Improved Distance and Mass Estimate for Sgr A* from a Multistar Orbit Analysis." *The Astrophysical Journal*. 830:1 (July 19, 2016).
Clayton, John. "The Soft Anthropic Principle." Does God Exist, Program 6. http://doesgodexist.org, 2013.
Cleaves, H. et al. "A Reassessment of Prebiotic Organic Synthesis in Neutral Planetary Atmospheres." *Origins of Life and Evolution of Biospheres*. 38:2 (April 2008): 105–115.
Choi, Charles Q. "Giant Galaxies May Be Better Cradles." http://space.com, 21 August 2015.
Christopher J. C. et al. "The Evolution of Galaxy Number Density at z < 8 and its Implications." *The Astrophysical Journal*. 830:2 (2016): 83.
Dole, Stephen H. *Habitable Planets for Man*. New York: Blaisdell, 1964.
Fernández, Yanga R. "The Nucleus of Comet Hale-Bopp (C/1995 O1): Size and Activity." *Earth, Moon, and Planets* 89:3 (2000).
Foing, Bernard, "If we had no Moon." *Astrobiology Magazine* https://www.astrobio.net/retrospections/if-we-had-no-moon/amp/, October 29, 2007.
Fountain, Henry. "Two Trillion Galaxies, at the Very Least." *The New York Times*. Last accessed October 17, 2016.
Gowanlock, M.G., D.R. Patton, and S.M. McConnell. "A Model of Habitability Within the Milky Way Galaxy." *Astrobiology*. 11:9 (2011): 855–873.

Hart, M. H. "Habitable zones about main sequence stars." *Icarus* 37:1 (1979): 351–357.

Holberg, J. *Sirius, Brightest Diamond in the Night Sky.* New York: Springer, 2007.

Huang, S. "Occurrence of life in the universe." *American Scientist* 47:3 (1959): 397–402.

Mackie, Glen. "To See the Universe in a Grain of Taranaki Sand." Centre for Astrophysics and Supercomputing. http://astronomy.swin.edu.au/~gmackie/billions.html, February 1, 2002. Appeared in *North and South Magazine* (New Zealand), May 1999.

Marsden, B. G. "Comet Shoemaker-Levy (1993e)." *IAU Circular*, 1993.

McMahan, David L. *The Making of Buddhist Modernism.* New York: Oxford University Press, 2008.

Meldrum A.N. "The Discovery of the Weight of the Air." *Nature* 78 (July 30, 1908).

NASA Blueshift. "How many stars in the sky?" https://asd.gsfc.nasa.gov/blueshift/index.php/2015/07/22/how-many-stars-in-the-milky-way/, January 22, 2015.

NASA. The NASA exoplanet archive. Exoplanet and candidate statistics. https://exoplanetarchive.ipac.caltech.edu/docs/counts_detail.html. Last accessed August 2018.

National Oceanographic and Atmospheric Administration. "Stratospheric Ozone." https://www.ozonelayer.noaa.gov/science/basics.htm, April 2018.

Paley, William. *Natural Theology: or, Evidences of the Existence and Attributes of the Deity* 1st ed. London: J. Faulder, 1802.

Penrose, R. *The Emperor's New Mind.* Oxford: Oxford University Press, 1989.

Phillips A.C. *The Physics of Stars*, 2nd Edition. New York: Wiley, 1999.

Reardon, Sara. "Haze Clears on Ancient Earth's Early Atmosphere." New Scientist. https://www.newscientist.com/article/dn21598-haze-clears-on-ancient-earths-early-atmosphere/, March 12, 2012. Accessed September 7, 2018.

Redd, Nola Taylor. "Spiral Galaxy Facts & Definition." https://www.space.com/19915-milky-way-galaxy.html, August 15, 2013.
Redd, Nola Taylor. "Red Giant Stars: Facts, Definition & the Future of the Sun." https://www.space.com/22471-red-giant-stars.html, August 21, 2013.
Sharpe, Tim. "Earth's Atmosphere: Composition, Climate & Weather." *Science and Astronomy.* https://www.space.com/17683-earth-atmosphere.html, October 13, 2017.
Scharping, Nathanial. "Why we shouldn't call exoplanets 'Earth-like' just yet." *Discover* (October 21, 2016).
_____. "Earth May Be a 1-in-700 Quintillion Kind of Place." *Discover* (February 22, 2016).
Zachrisson, E. et al. "Terrestrial planet across space and time." *The Astrophysical Journal* 833:2 (2016).

Chapter 6
Anitei, S. "The Smallest Genome: What's the Minimum DNA Amount for Life?" https://news.softpedia.com/news/The-Smallest-Genome-What-039-s-The-Minimum-DNA-Amount-for-Life-73763.shtml, December 12, 2007.
Alberts B., A. Johnson, J. Lewis, et al. *Molecular Biology of the Cell.* 4th edition. New York: Garland Science, 2002.
Axe, D. *Undeniable: How Biology Confirms Our Intuition That Life Is Designed.* New York: HarperCollins, 2017.
Bartel D.P., and J.W. Szostak. "Isolation of new ribozymes from a large pool of random sequences." *Science* 261 (1993): 1411–1418.
Chang, K. "*Life on Mars? Rover's Latest Discovery Puts It 'On the Table.'*" *The New York Times*, June 7, 2018
Chiaffrano, M. "Bartel and Szostak Experiment." https://www.slideshare.net/mchiaf36/biology-bartel-and-szostak-experiment, March 23, 2011. Last accessed June 1, 2018.
Crick, F. *Life Itself: Its Origin and Nature.* New York: Simon and Schuster, 1981.

Dalrymple, G. Brent. *The Age of the Earth*. Stanford, California: Stanford University Press, 1991.

Deamer, D. W. "The First Living Systems: a Bioenergetic Perspective." *Microbiology & Molecular Biology Reviews* 61:239 (1997).

———. "Calculating The Odds That Life Could Begin By Chance." Science 2.0. August 27, 2014. https://www.science20.com/stars_planets_life/calculating_odds_life_could_begin_chance. Last accessed May 28, 2018.

———. *First Life: Discovering the Connection between Stars, Cells, and How Life Began*. Berkeley, California: University of California Press, 2011.

Denton, Michael. *Evolution: A Theory in Crisis*. London: Burnett Books, 1985.

Eigenbrode, J. et al. "Organic matter preserved in 3-billion-year-old mudstones at Gale crater, Mars." *Science* 360:6393 (8 June 2018).

Fabrizio Cleri. *The physics of living systems*. Switzerland: Springer International, 2016.

Graham, Robert W. "Extraterrestrial Life in the Universe" (NASA Technical Memorandum 102363). Lewis Research Center. Cleveland, Ohio: NASA, February 1990.

Gutiérrez-Preciado, A., et al. "An Evolutionary Perspective on Amino Acids." *Nature Education* 3:9 (2010): 2.

Huber, H., et al. "A new phylum of Archaea represented by a nanosized hyperthermophilic symbiont." *Nature* 417 (2002): 63–67.

Kasting, J.F. "Earth's early atmosphere." *Science* 259 (1993).

Lehman, N. "RNA Self-assembly: Cooperation at the Origins of Life." KITP Multicell13 Conference. University of California Santa Barbara, February 13, 2013.

Lincoln, Tracy A., and Gerald F. Joyce. "Self-Sustained Replication of an RNA Enzyme." *Science* 232:5918 (27 February 2009): 1229–1232.

Luskin, Casey. "Top Five Problems with Current Origin-of-Life Theories." *Evolution News* (17 Apr. 2017). https://evolutionnews.org/2012/12/top_five_probl/. Last accessed August 2018.

Meyer, Stephen C. *Signature in the Cell: DNA and the Evidence for Intelligent Design.* New York: HarperOne, 2009.

Miller, S. "A Production of Amino Acids under Possible Primitive Earth Conditions." *Science* 117 (May 15, 1953): 528–529.

Nelson, David L., and Michael M. Cox. "Nucleotides and Nucleic Acids." In *Lehninger Principles of Biochemistry*, 3rd ed. New York: Worth, 2000.

National Institutes of Health. Genetics Home Reference: Your Guide to Understanding Genetic Conditions. https://ghr.nlm.nih.gov/primer/basics/gene, 2018. Last accessed May 12, 2018

Noller, Harry F. "Evolution of protein synthesis from an RNA world." *Cold Spring Harbor Perspectives in Biology.* 4:4 (April 1, 2012).

A. C. Phillips, *The Physics of Stars*, 2nd Edition (Wiley, 1999).

Rappé, Michael S., et al. "Cultivation of the ubiquitous SAR11 marine bacterioplankton clade." *Nature* 418:6898 (2002): 630–633.

Schopf, J.W., A.B. Kudryavtsev, A.D. Czaja, and A.B. Tripathi. "Evidence of Archean life: Stromatolites and microfossils." *Precambrian Research* 158 (5 October 2007): 141–155.

Shapiro, Robert. "Astrobiology: Life's beginnings." *Nature* 476 (August 3, 2011): 30–31.

USGS, "Age of the Earth." U.S. Geological Survey. Archived from the original on 23 December 2005. https://pubs.usgs.gov/gip/geotime/age.html, 1997. Last accessed May 1, 2018.

Wells, Jonathan. "Icons of Evolution: Why Much of What We Teach About Evolution Is Wrong." Washington DC: Regnery, 2000.

Witzany, Guenther. "Crucial steps to life: From chemical reactions to code using agents." *Biosystems* 140 (February 2016): 49–57.

Zerkle Aubrey L., Mouterde Claire, Shawn Domagal-Goldman, James Farquhar, and Simon W. Poulton. "A bistable organic-rich atmosphere on the Neoarchaean Earth." *Nature Geoscience* 5:5 (May 2012): 359–363.

Chapter 7

Adler, Mortimer J. *How to Read a Book: The Classic Guide to Intelligent Reading*. New York: Touchstone, 1972.

Buchanan, George W. *To the Hebrews*. In *The Anchor Bible* Vol 36. New York: Doubleday, 1972.

Copan P., and D. Jacoby. *Origins: The Ancient Impact and Modern Implications of Genesis 1–11*. New York: Morgan James, 2019.

Ehrman, Bart D. *Jesus: Apocalyptic Prophet of the New Millennium*. New York: Oxford University Press, 1999.

Fee, Gordon D. and Douglas Stuart. *How to Read the Bible for All Its Worth*. Grand Rapids, Michigan: Zondervan, 2003.

_____ and _____. *How to Read the Bible Book by Book: A Guided Tour*. Grand Rapids, Michigan: Eerdmans, 2002.

Fox, Evert. *The Five Books of Moses*. In *The Schocken Bible*, Volume 1. New York: Random House, 2000.

Gier, N. F. *God, Reason, and the Evangelicals*. Lanham, Maryland: University Press of America, 1987

Hill, Carol A. "Making Sense of the Numbers in Genesis." In *Perspectives on Science and Christian Faith. Journal of the American Scientific Affiliation* 55 (December 2003): No. 4.

Jones, Alexander Raymond. "Hipparchus." *Encyclopædia Britannica*, May 11, 2017. https://www.britannica.com/biography/Hipparchus-Greek-astronomer. Accessed August 10, 2018.

Hyers, Conrad. "The narrative form of Genesis 1: Cosmogonic, Yes; Scientific, No." *Journal of the American Scientific Affiliation* 36:4 (1984).

Jacoby, Douglas. *A Quick Overview of the Bible: How All the Pieces Fit Together*. Eugene, Oregon: Harvest House, 2012.

_____. *How We Got the Bible*. MP3 CD – CD, 2005

_____. Q & A 0294: The circle of the Earth, July 16, 2003. https://www.douglasjacoby.com/q-a-0294-the-circle-of-the-earth/

Lane W. R. "The Initiation of Creation." *Vetus Testamentum* Vol. 13, Fasc. 1 (January 1963).

Lightfoot, Neil R. *How We Got the Bible*. Grand Rapids: Baker Book House, 1988.

Longman, Tremper III, and John H. Walton. *The Lost World of the Flood: Mythology, Theology, and the Deluge Debate*. Downers Grove, Illinois: IVP Academic, 2017.

Oakes, John. *Is there a God? Questions and Answers about Science and the Bible*. Spring, Texas: Illumination, 2006.

Pilt, Peter. "Compelling Evidence of the Authenticity of the Bible." https://peterpilt.org, May 24, 2012. Last accessed November 2018.

Seely, Paul H. "The Firmament and the Water Above." *Westminster Theological Journal* 53 (1991).

Squire, Charles. *Celtic Myth and Legend*. Mineola, New York: Dover, 2003.

Walton, John H. *Genesis*. NIVAC. Grand Rapids, Michigan: Zondervan, 2001.

Chapter 8

Bahnsen G. L. *"The Inerrancy of the Autographs."* In *Inerrancy*. Edited by Norman L. Geisler. Grand Rapids, Michigan: Zondervan, 1980.

Bleibtreu, E. "Grisly Assyrian Record of Torture and Death." *Biblical Archaeology Review* 17:1 (Jan/Feb 1991).

Bloch, Iwan. *The Sexual Life of Our Time in Its Relations to Modern Civilization*. London: Rebman, 1909.

Blomberg, Craig L. *Historical Reliability of the Gospels*. Downers Grove, Illinois: InterVarsity, 1986.

_____. *The Historical Reliability of the Gospels*. Downers Grove, Illinois: InterVarsity, 2007.

_____. *Jesus and the Gospels: An Introduction and Survey* 2nd ed. Nashville, Tennessee: Broadman, 2009.

Bonhoeffer, Dietrich. Letter to Eberhard Bethge, 29 May 1944, 310–312 in *Letters and Papers from Prison*. Edited by Eberhard Bethge. Translated by Reginald H. Fuller. New York: Touchstone, 1997. Translation of Widerstand und Ergebung Munich: Christian Kaiser Verlag, 1970.

Bratcher, Dennis. "Word of Faith and "Commanding" Contextual Analysis of Matthew 21:21." The Synoptic Problem: The Literary Relationship of Matthew, Mark, and Luke. Accessed August 03, 2018. http://www.crivoice.org/commanding.html.

Bruce, F.F. *Second Thoughts on the Dead Sea Scrolls*. London: Paternoster, 1956.
Bunnin, Nicholas; Yu, Jiyuan, eds. (2004). *The Blackwell Dictionary of Western Philosophy*. Oxford: Blackwell, 2004.
Burrows, Millar. *The Dead Sea Scrolls*. Chicago: Moody, 1986.
Campus Reset. "Scientific Proof." April 10, 2017. Accessed September 02, 2018. http://www.campusreset.org/scientific-proof/, 2018.
Carrier, Richard Lane. *On the Historicity of Jesus: Why We Might Have Reason for Doubt*. Sheffield, UK: Sheffield Phoenix, 2014.
Charlesworth, James H. *The Historical Jesus: An Essential Guide*. Nashville, Tennessee: Abingdon, 2008.
Chisholm, Hugh, ed. "Celsus." *Encyclopedia Britannica* (11th ed.). Cambridge, Massachusetts: Cambridge University Press, 1911.
Cohen, Jennie. "6 Things You May Not Know About the Dead Sea Scrolls." https://www.history.com/news/6-things-you-may-not-know-about-the-dead-sea-scrolls, 7 May, 2013.
Coogan, Michael. *A Brief Introduction to the Old Testament*. Oxford: Oxford University Press, 2009, 257.
Copan P. and Jacoby D. *Origins: The Ancient Impact and Modern Implications of Genesis 1–11*. New York: Morgan James, 2019.
Curtis, Ken. "Whatever happened to the Twelve Apostles?" https://www.christianity.com/church/church-history/timeline/1-300/whatever-happened-to-the-twelve-apostles-11629558.html, April 10, 2010.
Dixon, Thomas. *Science and Religion: A Very Short Introduction*. New York: Oxford University Press, 2008.
Drummond, Henry. *The Ascent of Man*. New York: J. Pott, 1904.
Dunn, James, D.G. *Jesus Remembered*. Grand Rapids, Michigan: Eerdmans, 2003.
Eddy, Paul Rhodes, and Gregory A. Boyd. *The Jesus Legend: A Case for the Historical Reliability of the Synoptic Jesus Tradition*. Grand Rapids, Michigan: Baker Academic, 2008.
Ehrman, Bart D. *The New Testament: A Historical Introduction to the Early Christian Writings*. Oxford: Oxford University Press, 2008.

_____. *Misquoting Jesus: The Story Behind Who Changed the Bible and Why*. HarperSanFrancisco, 2005.

_____. *Did Jesus Exist? The Historical Argument for Jesus of Nazareth*. New York: HarperOne, 2012.

Evans, Craig. *Guide to the Dead Sea Scrolls*. Nashville, Tennessee: B&H, 2010.

Ferguson, Sinclair B; David F Wright; J. I. Packer. *New Dictionary of Theology*. Downers Grove, Illinois: InterVarsity. 1988.

Gathercole, Dr Simon. "What Is the Historical Evidence That Jesus Christ Lived and Died?" *The Guardian*. April 14, 2017. https://www.theguardian.com/world/2017/apr/14/what-is-the-historical-evidence-that-jesus-christ-lived-and-died.

Geisler, Norman and Nix, William. *A General Introduction to the Bible*. Chicago: Moody, 1986.

Gordon D. Fee. "The Textual Criticism of the New Testament." In Introductory Articles, Vol.1 of *The Expositor's Bible Commentary* ed. Frank E. Gaebelein. Grand Rapids, Michigan: Zondervan, 1979.

_____ and Douglas K Stuart. *How to Read the Bible for All Its Worth: A Guide to Understanding the Bible*. Grand Rapids, Michigan: Zondervan, 1982.

Grossman, Maxine. *Rediscovering the Dead Sea Scrolls: An Assessment of Old and New Approaches and Method*. Grand Rapids, Michigan: Eerdmans, 2010.

Hertzenberg, Stephanie. "6 Archaeological Discoveries that Support the Bible: The Holy Land is littered with evidence of biblical events." http://www.beliefnet.com/faiths/christianity/6-archaeological-discoveries-that-support-the-bible.aspx#dbGAvRdEbQpfwpca.99. Last accessed September 27, 2018.

Israel Museum Jerusalem. "The Digital Dead Sea Scrolls: Nature and Significance." http://dss.collections.imj.org.il/significance. Last accessed October 13, 2014.

Jacoby, D. *Compelling Evidence for God and the Bible: Finding Truth in an Age of Doubt*. Eugene, Oregon: Harvest House, 2010.

Josephus, F. *The Testimonium Flavianum*, Book 18, Chapter 3, point 3 (circa 71 CE), based on the translation of Louis H. Feldman.

The Loeb Classical Library.
http://www.josephus.org/testimonium.htm.
Kugler, Robert and Hartin, Patrick J. *An Introduction to the Bible*. Grand Rapids, Michigan: Eerdmans, 2009.
Layard, A.H. *The History of Assyria in the Ruins of Nineveh*. Royal Asiatic Society. London: John W. Parker, 1852.
_____. *Discoveries Among the Ruins of Nineveh and Babylon*. New York: G.P. Putnam, 1853.
Leslie, John. *Universes*. London: Routledge, 1989.
Levy, Leon. "The Digital Library: Introduction." Dead Sea Scrolls Digital Library. Last accessed October 13, 2014.
Louis, G., *Understanding History: A Primer of Historical Method*, 2d ed. New York: Alfred A. Knopf, 1969.
Lucian of Samosata. *The Death of Peregrine 11–13* (circa 165–175AD), translated by H.W. Fowler and F.G. Fowler in The Works of Lucian of Samosata, Vol 4. Oxford: Clarendon, 1949, as quoted and cited by Gary R. Habermas, *The Historical Jesus: Ancient Evidence for the Life of Christ*. Joplin, MO: College Press, 1996, 2008.
McKenzie, John L. *The Dictionary of the Bible*. New York: Simon and Schuster, 1995.
Mazar, Amihay. *Archaeology of the Land of the Bible, 10,000–586 B.C.E.* New York: Doubleday, 1990.
McDowell J., *Evidence That Demands a Verdict*, Revised Ed. San Bernardino, California: Here's Life, 1979.
Meier, John P. *A Marginal Jew, Vol. II*. New York: Doubleday, 1994.
Metzger, Bruce M. *The Text of the New Testament: Its Transmission, Corruption, and Restoration*. New York: Oxford University Press, 1964
Opus Dei, "Christian life: What do Roman and Jewish sources tell us about Jesus?" https://opusdei.org/en-us/article/what-do-roman-and-jewish-sources-tell-us-about-jesus/. Last accessed October 2, 2018.
Page, Stephanie. "A Stela of Adad-nirari III and Nergal-eres from Tell al Rimah." *Iraq* 30, 1968.
Perry, Simon. Resurrecting Interpretation. Bristol Baptist College. University of Bristol, 2005.

Pilt, Peter. "Compelling Evidence of the Authenticity of the Bible." https://peterpilt.org/2012/05/24/compelling-evidence-of-the-authenticity-of-the-bible/, May 24, 2012.

Rendsburg, Gary. "The biblical flood story in the light of the Gilgamesh flood account." In *Gilgamesh and the World of Assyria*, eds. Azize, J and N. Weeks. Leuven, Belgium: Peeters, 2007.

Tel Dan Inscription, The: "The First Historical Evidence of King David from the Bible." The Biblical Archaeology Society. https://www.biblicalarchaeology.org/daily/biblical-artifacts/the-tel-dan-inscription-the-first-historical-evidence-of-the-king-david-bible-story. Last accessed August 2018.

Thomas, Harry F. "How the Dead Sea Scrolls Were Found." *Biblical Archaeology Review* 1:4, December 1975.

Thompson, Thomas L. *Biblical Narrative and Palestine's History: Changing Perspectives 2*. London: Routledge, 2014

Trever, J.C. *The Dead Sea Scrolls*. Piscataway, NJ: Gorgias, 2003.

USGS, "Exploring the Deep Ocean Floor, This Dynamic Earth." USGS Publications Warehouse, 1999. https://pubs.usgs.gov/gip/dynamic/exploring.html, June 24,1999. Last accessed August 15, 2018.

Van Voorst, Robert E. *Jesus Outside the New Testament: An Introduction to the Ancient Evidence*. Grand Rapids, Michigan: Eerdmans, 2000.

Wallace, J. Warner. *Cold Case Christianity: A Homicide Detective Investigates the Claims of the Gospels*. Colorado Springs: David C. Cook, 2012.

Wilkenfeld, Wendy Anne. "Food Regulation in Biblical Law." Digital Access to Scholarship at Harvard. https://dash.harvard.edu/bitstream/handle/1/8846735/wwilkenfeld.html?sequence=2. Last accessed August 13, 2018.

Würthwein E. *The Text of the Old Testament: An Introduction to the Biblia Hebraica*. Grand Rapids, Michigan: Eerdmans, 2014.

Zuck, Roy B. *Basic Bible Interpretation*. Colorado Springs: David C. Cook, 1991.

Chapter 9

BioLogos Foundation, The. "Are gaps in scientific knowledge evidence for God?" BioLogos. The BioLogos Foundation. Last accessed August 2018.

Bloch, Iwan. *The Sexual Life of Our Time in Its Relations to Modern Civilization*, translated from the sixth German edition by M. Eden Paul M.D. London: Rebman 1909.

Bonhoeffer, Dietrich. "Letter to Eberhard Bethge, 29 May 1944." *Letters and Papers from Prison*. Edited by Eberhard Bethge. Translated by Reginald H. Fuller. New York: Touchstone, 1997.

Lewis C.S. *Mere Christianity*. New York: MacMillan, 1952.

Fischer, David Hackett. *Historians' Fallacies: Toward a Logic of Historical Thought*. New York: Harper & Row, 1970.

Flew, Antony. *A Dictionary of Philosophy: Revised Second Edition*. New York: Macmillan, 1984.

Grossman, Lisa. "Multiverse Gets Real with Glimpse of Big Bang Ripples." *New Scientist*. Special Report (March 18, 2014). https://www.newscientist.com/article/dn25249-multiverse-gets-real-with-glimpse-of-big-bang-ripples/. Accessed October 22, 2018.

Lewis, C.S. *The Problem of Pain*. New York: HarperCollins, 1940.

Zacharias, Ravi. *Jesus Among Other Gods*. Nashville, Tennessee: Thomas Nelson, 2000.

Made in the USA
Middletown, DE
11 May 2022